DUTCH LEANDER FRIGATE VAN SPEIJK

HMS Andromeda *(F-57) arriving in Den Helder in 1972.*
An improved Type 12 or Leander class General Purpose frigate. The
last warship built at HM Dockyard Portsmouth and commissioned on
2 December 1968. The first broad-beamed Leander.

VAN SPEIJK

Bottom: *An impression of the new frigate by C.A. Planten (1916-2003).*

INTRODUCTION

The British Leander class has been one of Britain's most successful warship designs with 26 units built for the Royal Navy, two for the Chilean Navy, six for the Indian Navy, two for the Royal New Zealand Navy, two units for the Royal Australian Navy's 'River' class and six for the Koninklijke Marine - the Van Speijk class.

Van Speijk class of the Royal Netherlands Navy emerged in the early 1960s from a need to replace the six Van Amstel (ex-US Cannon) class frigates. The British design was chosen in order to enable rapid construction. The Van Speijks were ordered in two batches; four in October 1962 and two in 1964. Orders were placed with the Nederlandsche Dok en Scheepsbouw Mij. in Amsterdam and Koninklijke Maatschappij De Schelde, Flushing, three ships each. All launched in a period of only two years - March 1965 to March 1967.

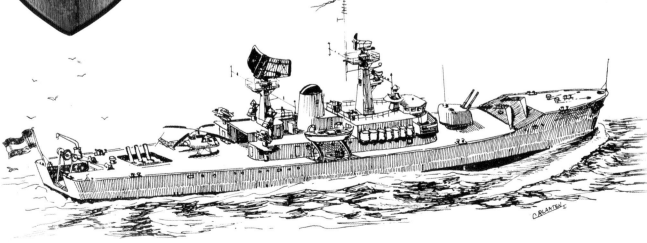

BRITISH LEANDER CLASS FRIGATES

The Leander Class was a development of the Type 12 frigate. In total 41 Type 12's, including 26 Leanders, were built for the Royal Navy. They formed the backbone of the post-war RN and were true maids of all work.

Type 12 refers to classes of the Royal Navy designed and built in the fifties and sixties until 1973 (*Ariadne* F 72):

1 The first Type 12 frigates, designed as 1st rate AS (convoy escorts), later named the Whitby class. Six operated in the Royal Navy, with one loaned to the Royal New Zealand Navy, and two built for the Indian Navy.

2 The design of the Type 12 Modified (Type 12M) or Rothesay class was optimised towards anti-submarine warfare and fleet escort duties. Fitted with the Seacat missile system. Nine were built for the Royal Navy, two for the Royal New Zealand Navy, and three (as the 'President class') for the South African Navy.

3 The third class, designed as general-purpose warship, was known as the Type 12 Improved (Type 12I) or Leander class. This class was made up of three 'batches'; the main differences between each batch being variations in propulsion machinery and weapons outfit. 26 were built for the Royal Navy, some of which later saw service in the navies of Chile, Ecuador, New Zealand, and Pakistan.

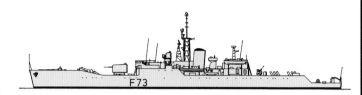

The Type 12 or Whitby class frigates were a six-ship class of anti-submarine frigates of the Royal Navy, which entered service late in the 1950s.

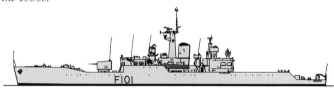

The Type 12M or Rothesay class. Twelve frigates were ordered, with the lead ship being laid down in 1956, two years after the last Whitby. The last three laid down were completed as improved Type 12 (Leander class).

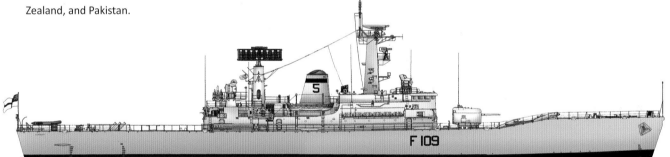

The 1963 edition of Jane's Fighting Ships described Leander class as a "mainly anti-submarine but flexible and all-purpose type"

Leander class (in 1963)	
Displacement	Tonnages: standard/full load 2,380/2860
Dimensions o.a.	*Length: 113.4 metres (372 ft)* Beam: 12.5 metres (41 ft), broad-beamed 13.1 metres (43 ft) Draught: 5.5 metres (18 ft) full load
Machinery	2 Babcock & Wilcox oil-fired boilers, geared steam turbines, 22,370 kilowatts (30,000 shp), 2 shafts
Max. Speed	29 kts
Complement	260
Armament	2 × 4.5-inch guns (1 × twin mounting Mk6) 1 × Seacat surface-to-air missile launcher 2 × 20mm guns (single mountings) 1 × Mk. 10 Limbo AS mortar

The Leander class have the same hull and substantially the same steam turbine machinery as the Whitby class, but are a revised and advanced design and did fulfil a composite anti-submarine, anti-aircraft and aircraft direction role.

The difference between the Leanders (Type 12I) and the Whitbys (Type 12) was most obviously that the stepped quarterdeck of the Type 12 had been done away with, resulting in a flush deck, with the exception of the raised forecastle. The superstructure had been combined into a single block amidships and a new bridge design gave improved visibility. A hangar and flight deck were provided aft for the Westland Wasp light anti-submarine helicopter, which was still at the prototype stage when the first ships were ordered. The ship was air conditioned throughout and there were no portholes in order to improve nuclear, biological and chemical defence. The ships were all given names which had previously been given to Royal Navy cruisers, mostly of characters from classical mythology, the exceptions being Cleopatra and Sirius.

HMS Andromeda *(F-57) arriving Den Helder in 1972. The Leander class was built in three batches between 1959 and 1973. It had an unusually high public profile, due to the popular BBC television drama series* Warship. *The* Leander *silhouette became synonymous with the Royal Navy from the 1960s until the 1980s.*

Centre:
The ships of the Van Speijk class were easy to recognize by their silhouette. The raised fore castle and their typical Signaal type of radar antenna made them quite distinctive

The Van Speijk class frigates originated from the English series of Leander class frigates. In detail, the layout was greatly changed, whereby the electrical and electronic part was entirely of Dutch origin and the fire control of the armament was partly of Dutch manufacture. Many smaller tools were also manufactured or built under license in the Netherlands. The hull construction, on the other hand, was largely unchanged and the propulsion system was also recreated in accordance with the English design. In view of the origin of the Van Speijk class frigates and their strong resemblance to the Leander class frigates (see Warship 2).

The Leander class frigates were the last ships in development from the Type 12 frigates with the accumulated sea experience of their predecessors already in service. They were also an important contributor to British export. A great deal of equipment was supplied for ships built abroad and a financial contribution was requested for the design. (For the Van Speijk class frigates, these license fees for the ship design amounted to £50,000 per ship and for the engine room design £20,000 per installation.)

Amsterdam 1978. HNLMS Isaac Sweers *(F 814),* Evertsen *(F 815) and HMS* Ariadne *(F 72). The ships were assigned to STANAVFORLANT, the NATO multi-national squadron. Note the differences between the frigates. (Coll. Jt. Mulder)*

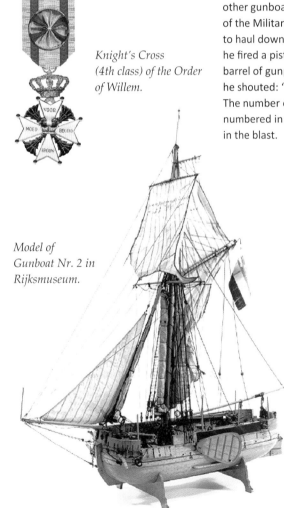

Jan Carel Josephus van Speyk
(31 January 1802 - 5 February 1831).
Miniature painted by J.A. Pluckx.

Knight's Cross
(4th class) of the Order
of Willem.

Model of
Gunboat Nr. 2 in
Rijksmuseum.

The name "Van Speijk" is a remembrance dedicated to a young officer.

The frigate was named after Jan Carel Josephus van Speyk (also written Van Speijk). He was a naval lieutenant who became a hero in the Netherlands for his opposition to the Belgian Revolution. Born in Amsterdam on 31 January 1802, Van Speijk was orphaned only a few weeks after his birth. When he was 18 years old, he joined the Royal Netherlands Navy and served in the Dutch East Indies from 1824 to 1828. He was involved in operations against illegal miners digging tin on the island of Bangka. This earned him the nickname "Scourge of the bandits".

When the Belgian rebellion commenced, he was commanding a gunboat. Van Speijk despised the Belgian independence movement, and he said he would rather die "than become an infamous Brabander". In October 1830 during "The bombardment of Antwerp" the beleaguered troops in the Citadel were supported by a squadron of gunboats. In November Lt. Van Speijk and some other gunboat captains were awarded the Knight's Cross (4th class) of the Military William Order. Belgians stormed his ship, demanding to haul down the Netherlands ensign. Rather than to surrender, he fired a pistol (some versions say he threw a lighted cigar) into a barrel of gunpowder in the ship's magazine. According to legend, he shouted: "*Dan liever de lucht in*" ("I'd rather be blown up"). The number of Belgians killed is unknown, though it probably numbered in the dozens. Twenty-eight of his 31 crew also perished in the blast.

Eight days after his death, the Netherlands declared a period of mourning. He was buried in the Nieuwe Kerk in Amsterdam, where the remains of Dutch naval hero Michiel de Ruyter are also interred. Van Speijk is regarded as naval hero in the Netherlands. This resulted in a Royal Decree (Koninklijk Besluit) number 81, 11 February 1831, issued by King William I pronouncing that as long as the Dutch Navy exists there will always be a ship named 'Van Speijk' to preserve his memory.[*] Legend has it that the mast of Van Speijk's ship is preserved at the 'Koninklijk Instituut voor de Marine' (Royal Netherlands Naval College).

A national memorial in his honour is located at the J.C.J. van Speijk lighthouse in Egmond aan Zee.

Page right:
Painting by J.J. Eeckhout and E.K.G. Wappers.

---

[*] In 1832 before the launch of the corvette *Argo* (800 tons) the name was changed in "*Van Speyk*".

Named *Van Speijk*

1 1832 - 1837

Corvette *Argo* (800 tons - 28 guns) was laid down in 1827. Name changed before the launch in 1832 in *Van Speijk*. Employed in the East Indies and transferred to the "Koloniale Marine" (regional navy) in 1837, new name *Medusa*. Acted for many years as a receiving/station ship at Surabaya.

2 1841 - 1878

Corvette *Medusa* (900 tons - 26 guns) laid down in 1838 also in Amsterdam. In 1841 allocated the name *Van Speijk* and launched in 1843. In 1868 transferred to the "Koloniale Marine", duties as receiving/station ship, stricken in 1878.

3 1882 - 1940

Screw steam vessel 1st class, wooden clad, iron hull, 3575 tons 2900 hp = 14 kts. Launched 1882; Between 1887 - 1897 *Van Speijk* made several terms to the East Indies. Dismantled; in 1904 adapted and furnished as accommodation hulk for stokers in training until 1940.

ETERNAL NAME

A proposal by Prince Frederik (Willem Karel), the second son of the king, led to a proclamation.

By Royal Decree dated 11 January 1831 nr. 81:
There always will be a ship of our navy carrying the name *Van Speijk*

After the war gunboat K 3 was refurbished and classified as a frigate. Employed as West Indies Guardship.

4 (1940) 1946 - 1960

Captured Gunboats *K 1 - K 2 - K 3* served in the "Kriegsmarine". After the war *K 3* was commissioned as frigate *Van Speijk*. Sold for scrap in 1960.

5 1960 - 1965

Accommodation vessel *Van Speijk* (ex-*Flores*), in Vlissingen.

6 1965 - 1986

Frigate and subject of Warship 14

Moored alongside jetty 4 Den Helder.
Flaghoist = Code (Interco)
Q D 1 = My engines are turning ahead.

7 1986 - 1994

Former minesweeper/hunter *Dokkum* decommissioned in 1983. In 1985 experimental vessel for diesel research (burning low grade fuels). Allocated visual callsign Y 8001.

8 1994 - 2021

Van Speijk (F 828) last M-(multi purpose) frigate of class of 8 ships. Full load 3320 tons- L.: 122.3 m (401.1 ft) L.: 1994 Comm. 1995 Modernized 2011 (new mast). As West Indies guardship scored many tons of cocaine. Laid up in 2021 due to staff shortage.

DESIGN AND PLANNING

On the slipway the first two units Van Speijk *(left) and* Tjerk Hiddes *in Amsterdam-North.* (NIMH)

In August 1959 representatives of the Netherlands Navy visited the Director General Ships in Bath for an exploratory conference about *Leander* class frigates. In this meeting a timetable was set for preliminary studies, producing design plans and specifications. In view of the need for timely replacement of aging destroyers and frigates it was decided to build a series of new frigates. To avoid a long lead time and also to take advantage of the adoption of this successful design.

Contracts for four ships were signed 15 January and for another two on 30 December 1963. The cost of each unit amounted about 30 million Dutch Guilders. Delivery of the first ship was set for 1 September 1965 but was delayed to 29 August 1966. Both yards argued they owed the delay to the slow output of the drawing rooms with design specifications arriving overdue in the yard.

Plans

The prototype for Dutch Leanders was *Arethusa*, laid down in September 1962.
Building plans and specifications were ordered from the shipyard J. Samuel White at Cowes, Isle of Wight.

- English measures were adapted and translated in the metric system.
- The compartments inside the ship would be arranged in a different way.
- Modified bridge for improved (always important) view.

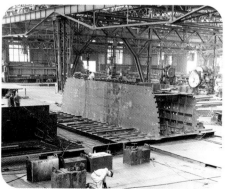

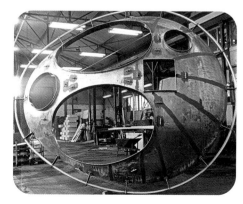

General management of the project was carried out by the NDSM.

For the mechanical equipment, this was carried out by the KMS. N.V. Groeneveld, van der Poll & Co (GROENPOL). was appointed to fit electrical installation in Amsterdam and N.V. van Rietschoten & Houwens in Rotterdam with N.V. Verebus in The Hague as the joint drawing office. Among the many important subcontractors, one can mention Bronswerk N.V. in Amersfoort (air supply and ventilation) and Van der Heem (communication equipment).

NDSM

The Nederlandsche Dok en Scheepsbouw Maatschappij (NDSM or Netherlands Dock and Shipbuilding Company), was a shipbuilding and repair yard based in Amsterdam, existing from 1946 to 1979. In the years 1965-1967 three frigates, *Van Speijk*, *Tjerk Hiddes* and *Isaac Sweers,* were completed.

Yard number 517 (Van Speijk) was launched on 5 March 1965. (NIMH)

As mentioned, the original British design of the Director General Ships in Bath was largely adapted by the 'Scheepsbouw Bureau' to merge Dutch insights in conjunction with the other technical departments. As a result, far-reaching modifications were made.

For example, an enlarged mess, and accommodation had been furnished in accordance with RNN requirements (light metal furniture). A bakery was added to the galley complex. A combined wheelhouse, gyro compass room and command shelter. The navigation bridge was enlarged, much Dutch equipment was installed. Air conditioning capacity was greatly increased. The engine room was British with Y 136 mod. machinery. Electrical installation was delivered following Dutch design. Radio and radar equipment were made in the Netherlands, as were the aiming devices and fire control. Sonar equipment was partly of American origin. An essential change was the arrangement of two Seacat SAM launchers on the hangar top deck instead of one.

Leanders showed a balanced internal division between operational, logistic and living spaces.

The diesel generators were installed in a separate watertight room, with a switchboard, as well as the cooling installation for the air supply. The converters were also placed low in a separate compartment.

The 4.5-inch Vickers gun was positioned at 30% of the length from the bow, with the ammunition magazines below. Near the stern the mortar was placed in a well with control and ammunition space. The close range missile system was located on the port and starboard hangar deck, with reload stores underneath. The main storage area was further back.

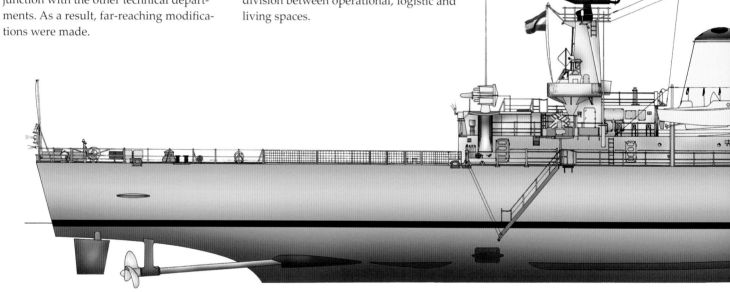

Van Speijk class Frigate						
Name	Pennant	Builder	Laid down	Launched	Commissioned	Fate
Van Speijk	F802	NDSM, Amsterdam	1 October 1963	5 March 1965	14 February 1967	Sold to Indonesia in 1986 as Slamet Riyadi (352)
Van Galen	F803	KM de Schelde, Vlissingen	25 July 1963	19 June 1965	1 March 1967	Sold to Indonesia in 1987 as Yos Sudarso (353),
Tjerk Hiddes	F804	NDSM, Amsterdam	1 June 1964	17 December 1965	16 August 1967	Sold to Indonesia in 1986 as Ahmad Yani (351)
Van Nes	F805	KM de Schelde, Vlissingen	25 July 1965	26 March 1966	9 August 1967	Sold to Indonesia in 1988 as Oswald Siahaan (354)
Isaac Sweers	F814	NDSM, Amsterdam	6 May 1965	10 March 1967	15 May 1968	Sold to Indonesia in 1990 as Karel Satsuitubun (356)
Evertsen	F815	KM de Schelde, Vlissingen	6 July 1965	18 June 1966	21 December 1967	Sold to Indonesia in 1989 as Abdul Halim Perdanakusuma (355)

*Hangar on display
at Navy Days.*
(Coll. Jt. Mulder)

Technical data			
Displacement:	2,200 tons standard, 2,850 tons full load		
Length:	113.4 m (372 ft 1 in)		
Beam:	12.5 m (41 ft 0 in)		
Draught:	4.2 m (14 ft 0 in) - max. 5.5 m (18 ft) at the screws		
Machinery:	2 x geared steam turbines 22,370 kW (30,000 shp)		
Speed:	2 shafts, 28.5 kts maximum / 12 kts cruising		
Range:	4,500 nm (8,300 km) at cruising speed		
	1967	**1979**	**2008 (Indonesia)**
Complement:	254	180	180
Armament:	2 × 4.5-inch guns twin mounting Mk6 2 × Seacat SAM launcher 1 × Mk.10 Limbo AS mortar	1 × OTO Melara 76 mm gun 2 × Seacat SAM launcher 8 x Harpoon anti-ship missiles 2 x Mk 32 torpedo launchers	1 × OTO Melara 76 mm gun 2 x Simbad twin launcher for Mistral IR SAM's 2 x 2 - C-802 SSM 2 x Mk 32 torpedo launchers
Helicopter:	Westland Wasp A.S. Mk.1	Westland WG-13 Lynx HAS Mk.2	MBB-105CB

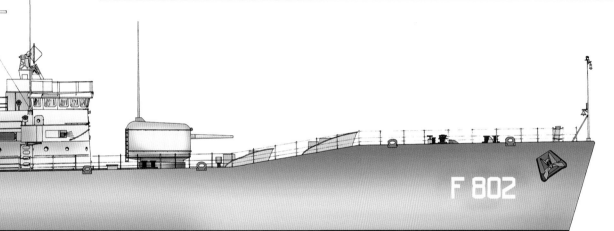

F 802

RAL 7038   RAL 7024

Profile drawing of Van Speijk.

Model plans

Plans are available at:
1- Netherlands Ministry of Defence:
   www.defensie.nl/onderwerpen/
   modelbouwtekeningen
2- NVM (Neth. Modellers Association):
   www.modelbouwtekeningen.nl

To avoid damage the screws were mounted in a drydock after launching.

The stern of Van Speijk *was fitted for the VDS,*

The VDS hoisting installation was housed in a well astern with doors as protection against water impact. [1]

The ops / sonar room [2], radio compartments and message handling office were on the starboard side of the H-deck. Damage control centre was also located in this area. The navigation bridge had a large setup offering a good all-around view. The instruments were concealed in panels against the bulkhead, it gave a large, neat and calm impression. It was slightly elevated in relation to the F-deck. Beneath the bridge was storage for instruments / tools / spare parts of the chart and ops-room and signal brigade.

It was well designed for the issue of meals. There was also a dishwashing machine with food waste disposer. In the evenings, the cafeteria could be divided by a sliding blind in a part for consumption (ship's canteen) while the other half for cinema or TV. The laundry [5] was located astern on the starboard side, opposite the cobblers workshop. The CPO's mess with pantry was on the starboard side.

Bunks.

On the H-deck forward portside was a small barber saloon next to the tailor shop. The accommodation concept was limited and in some areas even minimal. The complete crewlist counted 253 men!

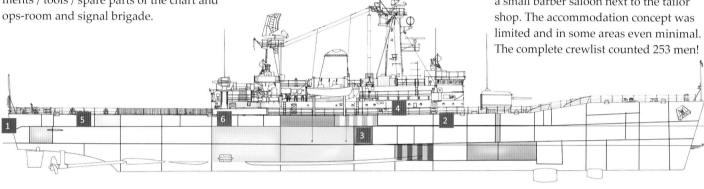

The bridge wings were located lower, so as not to impede the view through the bridge windows. The helmsman was on the J-deck [3] in a separate wheelhouse.

The crew was accommodated in quarters on the J and H decks, non-commissioned officers in cabins on the H and G decks and officers in 1- and 2-person cabins on the G-deck [4] and portside of E-deck. The location of the mess of the corporals (PO's) and ratings, the cafeteria, was just after the galley / bakery.

The wardroom and pantry were located below the officers' cabins, portside of the H-deck and to starboard the officers galley. The cabins, wardroom, CPO's mess and the cafeteria were well furnished and equipped with TV sets.
The Sick Bay [6], was on starboard opposite the cafeteria and had four berths. Supervised by the PO Sick Berth Attendant. Often a doctor embarked (some sailors had a certificate 'Medical Assistant'). The wardroom served as the auxiliary operating theatre in wartime.

Bridge

Wardroom.

Cafeteria (mess).

Galley.

The first of the Dutch Leanders, NDSM Yard number 517, was laid down on 1 October 1963 and completed 29 August 1966. (35 months later). A good achievement to show the capacity of the shipyard. The average building time of most of the 26 British Leanders was about 30+ months. On 14 February 1967 HNlMS *Van Speijk* was commissioned in Amsterdam. RADM (later VADM) P. Cool (1911-1971) can be mentioned as an important protagonist of the *Van Speijk* class project.

The *Van Speijk* class enjoyed the benefit to implement new developments regarding operational equipment. Also the progress in tactical thinking and evaluations improved the ASW capability.

Maiden Trip

In September 1966 *Van Speijk* sailed south to execute the shipbuilder's trials cruise. Ports of call were Dakar and Lisboa. During a full speed trial lasting for 12 hours *Van Speijk* logged about 360 nm, a continuous speed of 30 knots!
The NDSM/Navy crew was elated! Improvements and refinements reduced the total (standard) weight with 130 tons (5%). E.g. the total length of electrical cables network 110 km instead of 160 km. Building time suffered delay. Drawings of FSA 25 (HMS *Arethusa*) were sold to the Netherlands (for 150,000 Dutch Guilders) by shipyard J. Samuel White; the Br. Admiralty and English Electric. To "translate"them in metric and other adaptions took much time. Both yards complained about the slow output of the drawing rooms. Time sheet of the first and last Dutch Leanders; to compare with the last of the British Leanders:

Name	LD	L	Compl.	Comm.
Van Speijk	01.10.63	05.03.65	29.08.66	14.02.67
Isaac Sweers	06.05.65	10.03.67	20.02.68	15.05.68
Ariadne	01.11.69	10.09.71	-	10.02.73

Above: "On the Run" for a RAS.
Flag Romeo close up, FC/Radar antennas all trained over STBD as protection against the steel rod of the line-throwing gun! (NIMH)

As mentioned the scheduled delivery date of the first unit was 1 September 1965. The delay could be explained plausibly. Many changes in the design and a sometimes failing output of drawing rooms. Wage rates had been increased by a mere 10% and this urged the yards to put in a claim. In 1966 a commission was appointed and deliberate lengthy discussions led to an agreement in December 1968. An award of 1,820,000 Dutch Guilders was transferred.

From 22 April to 23 June 1971, Van Galen (top) and Evertsen made a journey to the Black Sea and Mediterranean. Visiting Piraeus (30 Apr - 4 May), Constanza (7 to 10 May, Odessa (11 - 14 May), Istanbul (15 - 18 May), Naples (1 - 8 June) pictured berthed at La Valetta (15 - 18 June), (Coll. Jt. Mulder)

Standing Naval Force Atlantic (now Standing NATO Maritime Group) in Port Everglades, the seaport of Fort Lauderdale, Florida in January 1972. The bow of USS Charles F. Adams (DDG 2), HNLMS Evertsen (F 815), HMS Jupiter (F 60) showing the Type 199 VDS and the West German frigate Braunschweig (F 225). Note the modified stern of the Netherlands ship. (Coll. Jt. Mulder)

Stern details of Van Galen.

Rudders and propellors arrangement.

MLM

The *Van Speijk* class proved to be reliable and effective in service but their design pre-dated the introduction of anti-ship missiles and consequently they became increasingly underarmed compared to later designs. To redress this loss of fighting power and reduce running costs, all six ships were given an extensive half-life modernisation; the MLM (Mid-Life Modernisation).

Staff objectives and requirements were given in 1969; guidelines forewarded by the Chief of the Naval Staff. A working group was established on 19 February 1973 for drawing up the new staff requirements.

Removed inventory awaiting disposal.

MLM

The objective was threefold:
1 bring up combat power
2 reduce operating costs through standardization in line with the guided missile and standard frigates and through personnel-saving investments
3 replace systems that had failed in practice or had been outdated

An ambitious project that went much further than intensive maintenance. Almost rebuilding the ship. Because of the many weaponry technical issues, the Navy Yard was selected for the job. (Although at the time the shipbuilding industries were desperately looking for orders.) Also, the timetable did put much pressure on the workers. Initially seven months were needed for disassembling and removing items. (Besides the main turbines all was removed from the engine room.) Thereafter new construction started and some testing was done. In all it was the largest job the Navy Yard ever did in their long history. Besides the many changes the ships kept their total displacement, because of new electronics and reducing the crew. It resulted in more living space for the crew and improved accommodation. At the yard countless minor and practical problems had to be solved. For instance, the frigates only had one longitudinal corridor were all the items had to pass. Pulling cables, painting and workers all went through. This demanded careful planning.

After two years at the yard a new Van Speijk emerged on 3 January 1979. It was the result of a joint effort of different Navy branches.	
Branch	**project hours**
'Centrum voor Automatisering Wapen- en Commandosystemen (CAWCS)'	5,000 h
'Bewapeningswerkplaatsen (BW)'	70,000 h
'Marine Electronisch en Optisch Bedrijf (MEOB)'	70,000 h
Navy Yard (RW)	580,000 h

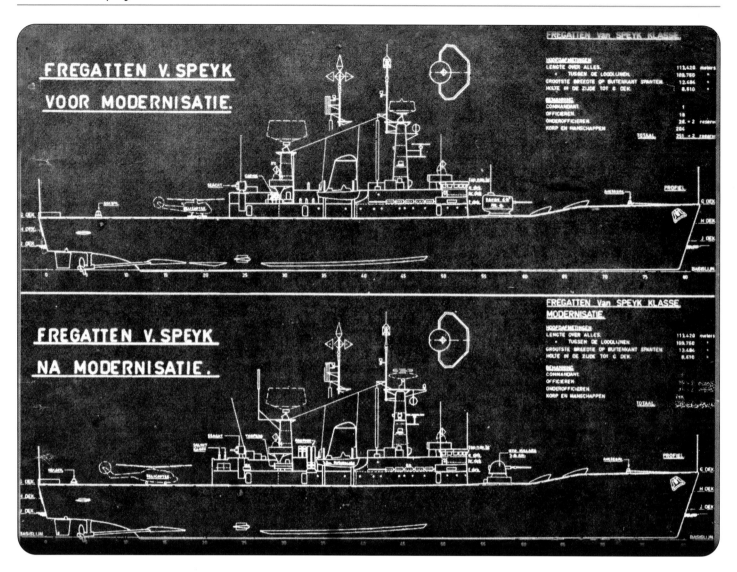

Mid-Life Modernisation. before and after.

DAISY

DIGITAL
> The 'Signaal Micromin Rekenaar' (SMR) the calculating brain of the system with 64K memory

AUTOMATIC
> The operator controls an array of automatic processes

INFORMATION PROCESSING
> The sensors forward their data to DAISY which processes an image on the screen. In certain situations the system warns the operator for danger and/or advices how to react.

SYSTEM
> The complete set off hardware. The software is stored on magnetic tapes.

Van Speijk had a M45-computer which assisted in fire control and ballistic math for the gun.

Most notable improvements made were:

- hangar and helicopter deck were adapted to operate the more capable Lynx helicopter,
- the twin 4.5-inch Mk.VI turret was replaced by the compact 76 mm rapid-fire Oto Melara gun,
- SSM Harpoon was mounted,
- the Mk 10 Limbo anti-submarine mortar was removed together with its associated sonar, and replaced by two triple Mk 32 anti-submarine torpedo tube mountings (STWS) being fitted on the weather deck to port and starboard of the hangar,
- after removing the Limbo the well was plated over to extend the flight deck,
- the OPS room was rearranged and improved,
- environmental hygiene improvements were made,
- the propulsion system received a major overhaul, including the introduction of full automation of the two Babcock and Wilcox boilers,
- the sensors were modified,
- updated communications equipment.

During official visits or special celebrations the ships can be illuminated (Van Nes and Van Galen) (Coll. H. Visser)

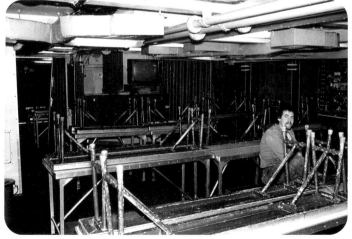

The main galley prepared breakfast, lunch and dinner for 250 men. It was situated adjacent to the dining hall for the junior ratings (a part for PO'). The Cafeteria system invited the men to walk alongside the counter to collect their meal. The meals for officers were served in the Wardroom, for chiefs in the CPO mess.

Cafeteria-mess deck has been swept and cleared up with a detergent. The mess deck sweeper waits until the deck is dry. To the left one can see the entrance to the PO-part of the dining hall.

Top: Impressions of MLM

Mounting the new gun.

Left and below: *Last stage of MLM.*

Mid-Life Modernisation		
Ship	from	to
Van Speijk	Dec 1976	Jan 1979
Van Galen	1977	1979
Van Nes	1978	1980
Tjerk Hiddes	1978	1981
Evertsen	1979	1982
Isaac Sweers	1980	1983

Van Galen passing.
Note: Funnel not yet modified, (Coll. J. Woort)

- DAISY was installed for command
- the electrical installation was improved
- living and working spaces were modernised
- a number of shipbuilding adjustments were made.

Due to increased automation due to the conversion the crew could be reduced from 253 to 180. In all it resulted in a significant decrease in running costs. The Operations Room was brought up to the standard of modern ships. OPS room staff and operators followed a secondary training.

In a joint effort the first ship was modernised in about two years. This was followed by executing a four months trials programme.

Den Helder, Nieuwe Haven, Jetty 17; Van Nes (F805) berthed, Van Galen (F803) is secured alongside. (Coll. Jt. Mulder)

Bows on Isaac Sweers,

A serious increase in anti-surface capability was achieved by fitting canisters for Harpoon surface-to-surface missiles close behind the funnel. Although intended to carry 8 of these canisters budgetary constraints meant that in practice normally only two Harpoon canisters were carried. Unfortunately it proved impossible within the limits of the budget and timescale available to replace the Seacat launchers with a more modern system but the electronics were updated. In a later stage all ships were fitted with a new infrared suppression funnel cap and *Evertsen* and *Isaac Sweers* were equipped with the US SQR 18A towed array sonar with its winch on the port quarter.

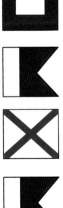

PAVA = International Call sign

Port Flag Locker *contained:*
- *26 alphabet flags*
- *10 cipher pennants*
- *4 substitute pennants*
- *10 cipher flags*
- *18 special flags & pennants*
- *1 Church pennant*

WEAPON SYSTEMS

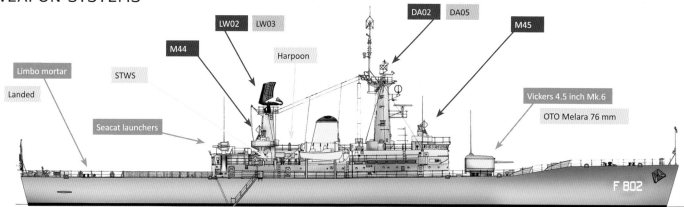

The Van Speijk class started their life as Netherlands built and sensored versions of the Leander class, carrying an extra Seacat missile launcher compared to their Royal Navy contemporaries. The Leander design dated from the fifties and twenty years on it was time to up-arm and boost combat capabilities. This was possible because this fine design offered room for installation of new and improved weapons.

In the RN a number had been converted from general purpose frigates to specialised vessels, armed with the Ikara anti-subma-rine weapon system (see Warship 2: HMS *Leander*) or with the Exocet surface-to-surface missiles. The *Van Speijk* class had been efficiently modified, not converted, they retained their general purpose capability.

GUNS

The original design incorporated several weapon systems that were copied. In 1967 this was a Vickers 4.5 inch Mk.6 Mod 3 twin turret. An early Cold War era dual purpose naval gun. Developed as a successor to former WW2 designs. Key improvements were higher rate of fire and anti-aircraft capability. The turret was power operated remotely in primary control but could also be locally operated in both power and hand control.

The guns were considered as accurate and produced a reasonably rapid rate of fire. Manual loading at 16 rpm/barrel was practice. Range was 16.4 km against surface targets and 7 km anti-air. For each barrel about 400 rounds were stored. When operating a crew of 26 (18 in turret) was needed. Among them many recollect the leaking hydraulic oils, some referred to the turret as Little Pernis (Oil terminals near Rotterdam). Exchanging the turret had serious consequences for the design. After all, the weaponry is heavy. When replacing in MLM for Oto Melara ballast had to be added as compensation for the lifted weight. After the turrets had been landed, they were sold for scrap.

In the fifties Vickers commenced to indicate guns by Arabic instead of Roman numerals. Sometimes the Mk.VI was referred to as Mk.6.

Weights
Vickers ammo:

Shell = 56 lb
Cartridge = 30 lb
Cordite = 12.5 lb

Technical data	Before MLM	Post MLM
	Vickers 4.5 inch Mk.6	OTO Melara 76mm
Designation	DP twin 4.5"/50 Mk. 6	76 mm/62 (3") Compact (updated)
Date in service	1946	1964
Calibre	4.5 inch / 11.4 cm	76 mm
Rate of fire	Designed: 20 rpm (power) Service: 12-14 rpm (manual)	Compact and Mark 75: 80 - 85 rpm (in automatic mode)
Max range (45°)	20,750 yards (18,970 m)	20,122 yards (18,400 m)
Shell weight	55 lbs. (25 kg)	13.88 lbs. (6.296 kg)
Train rate	25 degrees per second	60 degrees per second
Elevation	-15 / +80 degrees	-15 / +85 degrees
Train	About +150 / -150 degrees	Unlimited (uses a slip ring)
Total weight	45 tons	16,400 lbs. (7,439 kg) With ammunition and off-mount components: 18,783 lbs. (8,520 kg)

SAM (SURFACE-TO-AIR-MISSILE)
SEACAT GWS-20

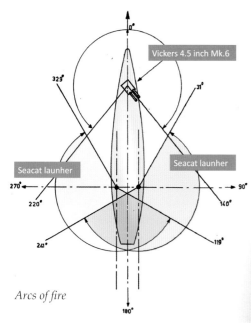

Arcs of fire

Designed, developed and manufactured by Short Bros & Harland Ltd, Belfast. In the 1960s the Seacat missile system was one of the most cost effective weapon systems ever developed. This first generation British missile was based on visual control of a guided weapon. A plain method of guiding a missile to its target. The operator tracked the target through a binocular sight guiding the missile by means of radio control, along the line of sight to the target.

Making the missile easier to track. Test results were first applied to the Australian Malkara anti-tank missile. In Great Britain the system was developed for a close range missile ship defence against air attacks. Designated as SX A5 test firings commenced in 1955 and within 18 months the concept was approved. After testing, the system was evaluated and in the spring of 1958 authorisation of SX A7 followed. The short development time can be explained by use of the anti-tank missile body. The system retained the thumb controlled joystick for its flight to target.

When the Royal Netherlands Navy ordered Seacat there were high expectations of the system. It was considered as a replacement for the 40 mm AA guns in use aboard the older ships. But also acting in the anti-ship role to destroy fast threats like MTB's at close range.

Initial research was directed towards air to air missiles, but soon it was realised that more sophisticated guidance systems would be needed for that purpose. However a more practical application was in the field of anti-tank weapons, where the targets would be either static or relatively slow moving, since the system worked best when the bearing of the target altered little during the time of flight.

The smaller size of the HSA M44 director was an advantage over the British model and enabled a second launcher on the hangar roof. Shortly after the order for Seacat projectiles was given the electronic industries exchanged tubes for transistors. A highly valued update.

Another problem were the HSA M44 / M45 director units which were not intended to be placed on open decks. Taking a severe punishment by wind and weather elements. It was redesigned and wind shields were mounted.

1982

During the Falklands War in 1982 about 20 Seacat systems were carried by RN ships. They gathered a poor score; a possible six kills. Most launches in combination with other air defence combatants. Seacat launches sighted by attackers caused violent, sometimes fatal, evasive action.

M40 series director

These directors used a roll and pitch stabilised enclosed dish antenna with conical scan for air target tracking (M44) and elliptical scan for surface targets (M45). Because the stabilizer was in the director, no below decks stable element (axis conversion) was needed. The initial data was fed by the detection radar, then the director automatically locked on and tracked. There was no optical range-finder, the director operator acquired and followed the target using a pair of binoculars on an aiming bar. The

view differed depending on application: 12 deg. (magnification 4x) for M44 Seacat control and 8 deg. (magnification 6x) for M45 gun control. The associated solid state digital computer of M45 was hard-wired for particular ballistics (for 2 or 3 different

weapons depending on use of memory). Guidance was generally visual, but the M44 could guide a missile by radar. Total weight of the director was 1,250 kg.

Top, right:
After MLM Seacat launchers and associated M44 were retained. Note the new hangar door.

Van Galen, *shows the general layout of the launcher deck and mast before MLM. The crew is preparing the ship for a port visit. The photo was taken from the Wasp helicopter, while the flight crew prepares for landing.*
Note the windshields around the M44 directors. (NIMH)

Flag India is c/u (close up) ; meaning: 'I am ready to receive you alongside'. Amidships port: two fenders are lowered.

MK.10 A/S MORTAR

Three-barrelled mortar Limbo was the final development of British World War II ahead throwing ASW weapons. It was designed to deal with a submarine moving faster than 15 kts. Range was controlled by opening gas vents and was between 400 and 1,000 yards (366 and 910 m). The barrels were aimed by a combination of roll and pitch compensators which also stabilised the mounting. The Mk.10 was designed such that the mortar bombs always entered the water at the same angle, which simplified the fire control solution.

It was mounted in a recessed well aft of the flight deck so that when the barrels were tilted over horizontally they were in line with the loading ports leading from the adjacent mortar handling room and magazine. A pneumatic system rammed the projectile into the barrels which could be raised to the firing position by electric

motors driving the mounting in the pitch and roll axes. The three barrels were adjustable for elevation by remote control so that range varied from a minimum of

around 290 yards up to the maximum. A salvo was spread in a triangular pattern around the target position so as to maximise the chance of a kill.

Each of the barrels was pivotted a third of the way up from the lower end, counterweights being fitted at the base to assist rapid movement in elevation. The barrels were interconnected and carried on a cradle which was free to rotate laterally on bearings at either end of the mounting structure. The whole installation was stabilised in roll and pitch by a metadyne system using reference data from the ships stable gyro platform.

Limbo was teamed with the search and attack sonars and a fire control suite. The gathered information was used to compute the speed, course, position and depth of the submarine. The firing angle was then calculated to allow for the time of flight of the mortar bombs as well as additional time taken to sink to the required depth. This information was constantly displayed by means of dials showing target parameters and the required angles for firing. The mounting was automatically kept aligned to the calculated impact point by remote control, firing being manual initiated.

After landing the Limbo of Van Speijk it was transported to the Navy Museum nearby. The red-white-blue ribbons on the barrels were not official but a decoration by the crew.

Roll motor

Roll gear train

Pitch receiver

Pitch motor brake housing

Cradle trunnion bearing

Refit with Harpoon and Mk. 46 torpedoes.

THE MK.32 TT AND MK.46 TORPEDOES

In the MLM refit Limbo was removed and the ship equipped with 'short' Mk.32 mod.5 torpedo tubes to launch the Honeywell Mk.46 mod.2 anti-submarine lightweight torpedo (active/passive homing). The first lightweight torpedo, the Mk.43, was introduced in 1951, followed by the Mk.44 in 1957. The third generation of Mk.46 was introduced in 1964. The mod.2 warhead provided 27% more explosive power over the mods. 0-1. Range 6 nm (11 km); homing depth 1500 ft. (450 m,) at 40 kts.

HARPOON

It took some time to choose a surface-to-surface missile to complement the Oto Melara. The main contenders were the French MM38 (Exocet) and the American

RGM-84 Harpoon. Eventually a decision was made in favour of the more advanced (and more costly) Harpoon, for which ramps can be seen just forward of the twin uptakes.

In 1971 McDonell Douglas was selected by the US Navy as prime contractor for

developing the missile. By August 1996, 6265 missiles were delivered to USN and foreign customers. The missile is a 'fire and forget' weapon, being provided with pre-launch target data, using inertial guidance techniques during its outward flight and finally its own radar seeker for terminal

Technical data	
	Harpoon
Designation	RGM-84
Date In Service	1977
Class	Subsonic Cruise Missile
Length	4.628 m
Diameter	0.343 m
Guidance	Inertial, semi-active radar
Speed	0.85 Mach (High subsonic)
Range	67 nm = 125 km
Weight	681 kg (224 kg warhead)

guidance. Making the missile independent of the ship's sensors. It cruises at low altitude directed by altimeter control. The active radar homing system switches on automatically at a predetermined distance from the position of the target at launch. The homing radar, which is frequency-agile to prevent jamming, then selects and locks on the target. In the final phase the missile executes a climb /dive manoeuvre. At a length of 4.63 m (shorter than the horizon-range Exocet) Harpoon was particularly compact for a SS missile. Harpoon is also a little slower, around 500 knots compared to 611 for the nearly Mach 1 Exocet. The price for which is a much

shorter range on the MM38 of about 27 nm in low and 43 nm in a higher altitude trajectory. Harpoon has gone through several development stages known as Blocks. The initial model is known as Block 1.

OPS Room

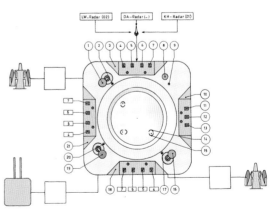

The top view of SEWACO control.

OPS room post MLM. The need for quick response on incoming aircraft led to automatisation. The senior OPS officer was appointed PWO (Principal Warfare Officer) and managed the entire SEWACO system. (RN had two PWOs: PWO-AIR and PWO-A/S.)

Both pages:
Fitting out Van Nes *at De Schelde 1966.*

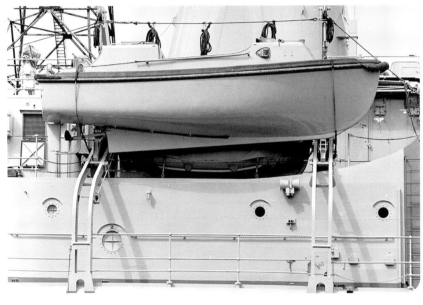

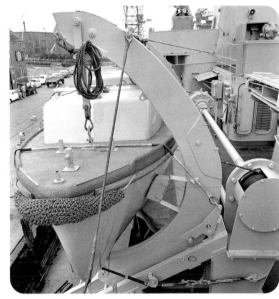

WESTLAND WASP

When the new frigates of the Van Speijk class entered service 12 Wasp helicopters had been ordered for the fleet air arm (MLD). The first two arrived in November 1966 and the last one in June 1967. This Manned Anti-Submarine Torpedo-Carrying Helicopter (MATCH*) acted as a weapon carrier to extend the range of the "naval armament". The Wasp had a high four-legged undercarriage with wheels for use aboard. Furthermore, the main rotor was equipped with "negative pitch": the angle of the blades could be adjusted downwards, so that the main rotor could press the helicopter against the deck, until lashed. The main rotor blades and tail could be folded to fit the helicopter in the small hangar.

The 12 helicopters (batch 2 and 2A, designated AH-12A) were stationed at De Kooy and assigned to VSQ 860. While embarked they flew mainly as torpedo weapon carrier (1 torpedo). After the ship established a contact the helicopter was directed to deliver a homing torpedo. Other duties included light transport and a rescue role.

Five of the helicopters had dual controls for flying practice. As the first on-board helicopter of the Navy, the Wasp proved to be adequate for its task, although its performance in warm weather was lacking. Then they often flew without doors and back seats to make the helicopter as light as possible. Some crew members called it 'A doghouse with roof engine' but many silently loved the little 'Pony'. Over the years three Wasp helicopters had to be written off.

Technical data	
Westland Wasp Mk.1 series 2 & 2A	
NL Designation	AH-12A
Date In Service	1966-67
Pennant	235 - 246, 247
Length	body 9.24 m / total 12.29
Rotor diameter	9.83 m
Engine	Bristol Siddeley Nimbus 503 turboshaft rated at 710 shp
Speed	193 km/h
Range	435 km
Weight /AUW*	2500 kg
Complement	2 (pilot and mecano) + 3 seats
* AUW = All-up Weight	

15 June 1967. Lt. Remmen executed the first landing on the flight deck of Van Speijk. Note the position of the wheels, here blocked for a save landing on land. Front wheels in line, rear wheels on 45 degrees so they could act as brake when sliding forward. At sea all 4 wheels would be in 45 degrees position. The option of only blocking front wheels, for running take off was not in use. (NIMH)

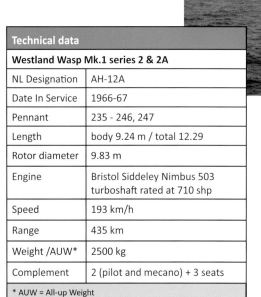

Pony 236, embarked on Isaac Sweers as weapon carrier with a torpedo. (NIMH)

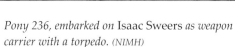

In 1974 a substitute was acquired. When in 1981 the Wasp was succeeded by the Lynx, 10 Wasps were transferred to Indonesia.

---

*   In NATO: Medium range Anti Submarine Torpedo Carrying Helicopter

WESTLAND WG 13 LYNX

While the ships received MLM the hangar and flight deck* had been modified for handling the much heavier Westand Lynx. In its ASW role the Lynx carried two American Mk 46 active/passive acoustic homing torpedoes. These could also be launched by the ship's torpedo tubes.

Technical data	
Westland WG 13 Lynx	
Designation	Westland Lynx SH-14B
Date in service	1979 - 1980
Pennant	266 - 275
Length	11.66 m / 15.16 m (overall)
Rotor diameter	12.8 m
Engine	2x RR Gem FP4RR-1010 of 1120 shp
Speed	270 km/h
Range	630 km (3.5 hours)
Weight/AUW	4765 kg
Complement	2 or 3

Crew depending helicopter & mission:
- Pilot
- 2nd pilot
- Tacco (Tactical Coordinator)
- Mech (Mechanician)
- Sonar operator
- Hoist operator/swimmer/ diver

- Six Mk.25 (UH-14A) received 1976-77 Land variant SAR/ rescue hoist-transport/utility
- Ten Mk.27 (SH-14B) received 1979-80 ASW variant Dipper/Alacatal sonar DUAV 4A
- Eight Mk.81 (SH-14C) received 1980-81 ASW variant MAD/Texas Instruments Magnetic Anomaly Detection

Note: As of 1993 STAMOL = Standardisation and Modernisation Lynx all helo's designated SH-14D. Stationed at NAS De Kooy, near Den Helder the helicopters were operated by the resident squadrons, nos. 7 and 860, the former being the training and SAR-unit, the latter the shipbased ASW and special missions unit. A hydraulic winch (hoist) could lift 600 Lb (270 kg) was an important asset to rescue shipwrecked sailors. Underslung loads up to 3000 Lb (1360 kg) could be carried on an external freight hook.

*   In Royal Netherlands Navy called 'heli dek'

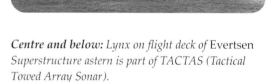

Centre and below: Lynx on flight deck of Evertsen *Superstructure astern is part of TACTAS (Tactical Towed Array Sonar).*

HSA radars	
Category	type
LW	Lucht Waarschuwing (long range surveillance)
ZW	Zee Waarschuwing (surface warning)
DA	Doelsaanwijzing (target acquisition and air search)
VI	Vertikale Interceptie (height-finding)

SENSORS AND ANTENNA

Like the older ASW destroyers the new frigates combined LW-02 with DA-02. (See page 24)

Long range radar LW02 / LW03
The LW02 (in MLM replaced by LW03) was the first post-war destroyer radar. It was stabilised and provided wide area air search. LW01/ -02/ -03/ -04 were also called SGR*-114, the long range early warning radars (wavelength 25 cm) They were capable of detecting aircraft at about 85 nm, although 100 nm was claimed. Antenna dimensions were 7.8 x 4.25 m (25.6 x 13.9 ft.) with a weight of 953 kg (2100 lbs). Compared to LW02, the LW03 had a higher rotation rate.

___
* SGR stands for Search Ground Radar

DA-02 / DA-05
The DA 01/ -02/ -04/ -05 were also called SGR-105 or the medium range warning and target indication radar (wavelength 10 centimeter). Improved version of the DA-01 carried by the ASW-destroyers (Warship 10). In MLM it was replaced by DA-05 which was the first of a new series of DA-radars. Much more powerful and better performance. Detection of a 2 m target at 72-84 nm and about 60,000 ft was claimed.

Left: LW02 antenna.
Note the helicopter glide path indicator halfway the mast.

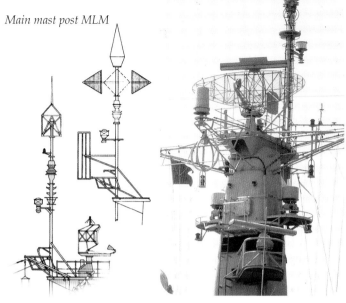

Main mast post MLM

Birdcage and Pointed Cap

Van Speijk (F 802), *Tjerk Hiddes* (F 804) and *Isaac Sweers* (F 814) were built at NDSM and equipped with the conspicuous birdcage antenna of the Bellini-Tosi system for the RACAL FH-5 direction finder. The frigates built at "De Schelde" were equipped with UA-13 in the higher frequency spectre, called "Puntmuts". The FH-5 operated in the frequency range between 30 kHz and 30 MHz, i.e. in the low, medium and high frequency bands LF/MF/HF.
The frequency range of the UA-13 covers the HF / VHF / UHF (high, very high, ultra high frequency) bands.

Flag BRAVO
meanings:
- *Weapon practices*
- *Fuelling or trans-*
 ferring explosives
- *Transporting*
 explosives or fuel

Van Nes post MLM

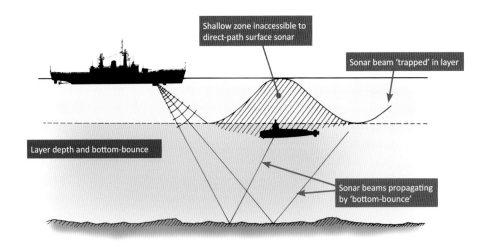

Shallow zone inaccessible to direct-path surface sonar

Sonar beam 'trapped' in layer

Layer depth and bottom-bounce

Sonar beams propagating by 'bottom-bounce'

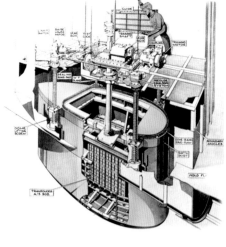

Exploded view of a Type 177 sonar (not on Van Speijk) shows the construction of the retractable dome.

SONARS

• CWE-610 a long range active/passive hull mounted sonar; max range depending bathy thermographic conditions 30 K yard.

• Sonar PCE-162. Detects/classifies mid water seabed targets. Displays port/stbd recordings simultaneously on a single recorder which has a max. range scale of 1200 yards.

The maximum working speed was about 20 knots. However, sonar echoes are very difficult to detect beyond 15 knots because the water rushing by the dome would mask the incoming echos. The dome was normally housed inside the hull for any of the following conditions: when working cables or bottom lines; when entering or leaving harbour; when steaming into a heavy sea; if cessation of sonar operations could be tolerated; and finally, when navigating in shoal waters and sonar was not required for navigation.

The Limbo mortar required three-dimensional information for operation which was obtained by PAE-170B sonar for search and fire control. It was developed to deal with fast (15 kts) diesel submarines capable of diving to 1000 ft. The Limbo depth settings were taken directly from the transducer's tilt angle. The 170B could also accept visual and radar data (e.g. the location of a submarine periscope). The production of the PAE-170B sonar started in 1954. It was removed when the Limbo was landed in MLM.

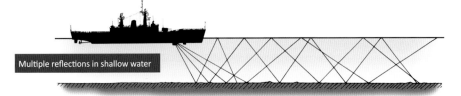

Multiple reflections in shallow water

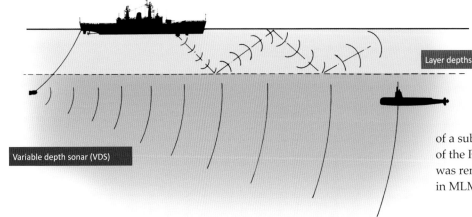

Layer depths

Variable depth sonar (VDS)

Edo salue les frégates de la classe «Van Speijk»

French language advertisement of the US 'EDO corporation', manufacturer of the VDS

VDS

Astern was the VDS (Variable Depth Sonar). This EDO 700 type was developed in parallel with the 610-series hull mounted sonar and designated PDE-700 (Ping Depth Escort). Streaming this device was hard work for the crew. Starting with unrolling the 300 meter, 10 cm diameter tube. It was floating and had to be connected to the 'fish'. Now the VDS was ready for streaming. Since most of the time submarine hunting was in the North Atlantic it was often a job in cold weather conditions.

After the exercise the VDS had to be disconnected again. Often in a dire situation, as the ship proceeded at slow speed to keep some tension on the tube.

Replaced ASW support equipment during MLM

• The towed BT = Bathythermograph by an X-BT = Expendable BT measuring temperatures at various depths hence identifying thermal layers. Subs could hide under

In MLM a new single sonar dome was fitted.

the layer and escape detection. Salinity and density of the sea are important aspects.

• The T-Mk.6 Fanfare torpedo decoy system by the NIXIE AN/SLQ25 passive electro acoustic torpedo decoy system.

• New Underwater Telephone ATM504A successor of the so called "Tuum"; for comms the Operational brevity code "Gertrude" was transmitted.

HNLMS Van Speijk

1967 - 1986

Commissioned	First foreign port call	Port visits	Most distant port
14 February 1967	February 1967 Portland	61 different foreign ports	Wellington (18.516,79 km)
Named *Van Speijk*	**First time Panama Canal**	**Most often visited port**	**"World Cruise 1970"**
6th ship	January 1970	Portland (5 times)	**Distance covered.** 40,833.7 nm (75,000 Km) nearly 2 times around the world
		Last foreign port call	**Fate**
		12-15 October 1985 Helsinki	Indonesian Navy 1986 - 2019 KRI *Slamet Riyadi* (352) Sunk offshore as dive attraction

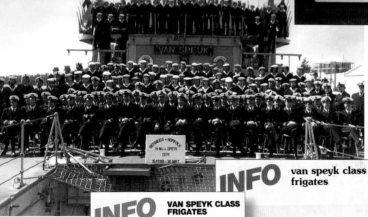

The crew of spring 1974/

Commanding Officers		
From	To	
1967	1968	Kltz J.H. Weijers
1968	1968	Ltz1 V. Esbach (sub.)
1968	1969	Kltz A.W. Crince le Roy
1969	1970	Kltz J.J. Valk
1970	1971	Kltz D.N. Wentholt
1972	1973	Kltz H.L. van Beek
1973	1973	Kltz P.J.A. Wehrens
1973	1975	Kltz H.C. van der Lee
1975	1976	Kltz P.A.A.J. van Oppen
1979	1980	Kltz H.A.J. Nijenhuis
1980	1982	Kltz P.W.F. Brunsmann
1982	1984	Kltz J.P.H. van Noort
1984	1985	Ltz1 H.A.W. Rötgers

OPERATIONAL HISTORY

Laid down on 1 October 1963 at the Nederlandsche Dok- en Scheepsbouw Maatschappij in Amsterdam. On 5 March 1965 with appropriate ceremony *Van Speijk* was launched by Mrs. A.H. van Es - Kat spouse of Deputy Minister of Defence (Navy), Rear Admiral A. van Es.*

Right and centre:
Not yet commissioned! On shakedown cruise in September 1966. (Dakar and Lisbon) Note: absence of pennant number.

On 14 February 1967, the frigate was commissioned by Commander J.H. Weijers with the usual ceremony. The frigate was commissioned as *Van Speijk* (F 802). Leaving port on 22 February for Portland, Gibraltar and Leixões. Entering Den Helder at 20 March.

On 3 April the second guarantee voyage commenced from Den Helder via Belfast, Swansea, Portsmouth and Rotterdam. Arrival Den Helder on 24 April.

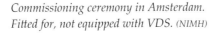

Commissioning ceremony in Amsterdam. Fitted for, not equipped with VDS. (NIMH)

* Dubbed 'Iron Adrian' by his contemporaries

Amsterdam, Tuesday 14 February 1967.
All Hands. A bright day, frosty wind. The crew mustered on
the heli deck, the Marine Band of the Royal Netherlands Navy
perched on top of the hangar. The first of a new class frigates
to be commissioned. Note: Shipbuilder's flag (NDSM) still
flying. (NIMH)

First NL frigate in FOST

On 5 May 1967 *Van Speijk* departed Den Helder bound for Portland.

Upon arrival a crowd of Sea Riders boarded, a warm welcome by the Staff of FOST (Flag Officer Sea Traing). Our opposite numbers visited every department. They would coach and help us during the Work-Up. The crew tried to reach a high standard of readiness and to improve operational capability.

The crew was eager to learn "FOST jargon" and studied the WPP (Weekly Practice Program) crammed with "serials". With enthusiasm one discussed about: Shakedown Week with Staff Sea Check, Harbour Training and Defects Week, Ceremonial Entry, Thursday War, lessons in HMS *Osprey*, Boarding a merchant ship, Chasing "the clockwork mouse" = ASW = AA and gunnery practice, Action Messing, RHOC (Return Harbour On Completion), Disaster relief party and Ship's Landing Party to name a few. Qualificatory expressions were: Below Standard, Satisfactory, Very Satisfactory and Good.

Van Speijk returned on 20 July. As one Flag Officer mentioned: "*Dutch, I expect you will do your utmost best! War can come very soon!*".

On 25 September *Van Speijk* joined the Neth. Task Group. Notable because for the first time a Dutch warship sailed with helicopter embarked. After some exercises the TG returned on 2 October.

Autumn cruise 1967

Departing 23 October, assigned to the Task Group: *Karel Doorman* (flag), *De Zeven Provinciën, Poolster, Drenthe, Van Nes, Noord Brabant, Van Speijk, Zeeland, Limburg, Potvis, Dolfijn* and *Walrus*. Several politicians and press were embarked to witness navy life on a frigate for a day. Demonstrating some weapon systems, underway replenishment and the Wasp helicopter while enjoying the splendid weather conditions. The guests would be ferried to Rotterdam later that day to disembark after which the ship joined the TG again, now steering a southerly MLA (Mean Line of Advance). When passing the Strait of Gibraltar on 29 October, the frigate entered the bay to collect the mail for the TG.

On 31 October the ships were south of Mallorca, where *Van Speijk* executed RAS with *Poolster*. Next day the ships found themselves in heavy seas caused by a strong Mistral. Nevertheless, they did not postpone the scheduled exercises. On 3 November the TG entered the port of Toulon where they berthed near the French navy complex l'Arsenal. The few days in port were used for small maintenance and leisure. The TG departed on 13 November for Barcelona. The next morning in close formation so public relations could get their requested pictures taken from two helicopters. Two days later the TG was split up. While a group visited Barcelona, the others, amongst them *Van Speijk*, went to Valencia (17 – 20 Nov). After assembling the TG, the ships set course for Gibraltar, passing the Strait of Gibraltar in the night of 24/25 November. On 26 November *Van Speijk* acted independently and proceeded to Den Helder for tilt definitions; to define and calculate the critical angle of list by rolling. The stabilizers with automatic control by gyroscope reduced rolling by one-third! The pointer of the clinometer seldom passed two ciphers on the dial.

29 Nov. The flagship anchored at 'Pompey' roads to collect VIP's and press. Next day *Van Speijk* joined the carrier in the North Sea at noon and after entertaining the guests *Van Speijk* moored at the homeport at 1800 hours. "Mission accomplished"…

Stability tests in Den Helder 1967. (Coll. H. Visser)

September 1967; Bay of Biscay, a Rougher.

Man overboard.
Beaufort 6 recently had been decreased but still a lively seastate (3-4). *Poolster* and *Van Speijk* executing a RAS with (as usual) a Romeo Corpen (replenishment course) against the wind to avoid yawing. Suddenly *Van Speijk* pitched in a head sea and the focsle was hit by a slammer. The able seaman with the distance line in his hand was swept overboard.

Marking position in the plot by the OPS room an emergency breakaway procedure (Emergency six). Then full astern was given and the duty rescue swimmer jumped from portside in the water. Ship manoeuvring to avoid the man overboard. This seaman inflated his lifejacket and recognised his mate, the rescue swimmer. Their position was about 30 yards abreast the bridge of *Van Speijk*. The able seaman could be retrieved by a small special constructed crane.

After changing dress, a mug of coffee and a fag in the sickbay he resumed his duties.

FOST Lessons learned !
****
5 May 1967 *Van Speijk* sailed to Portland to report for the Sea Training Work-Up
Returned 20 July. Summer leave 27 July – 22 August.
The FOST Staff/Seariders would be delighted when they learned about the rescue action by their "pupils".

September 1967:
Photos: (Leicacord) by LtCdr H.C.J. van der Woude, Chief Engineer
Poolster. *(Capt. T Jan.1973).*
(Coll. H. Visser)

*December 1967
Fitting the VDS
during guarantee
maintenance in
Amsterdam.*

The opponent watching Van Speijk. *A new Kresta I with Ka-25 Hormone.*
(Coll. H. Visser)

INTEL:

Monitoring & Report
- **Shadowing** = A ship could be designated to collect (electronic) info and observe behaviour/photos

Other code words used:
- **Tattletail** = Keep in contact to relay targetting info. Own attack not needed.
- **Marking** = Be ready to attack at once.

Note: Shining F/C (Fire Control) radars is a hostile act.

1968
Since 8 December 1967 *Van Speijk* had been returned to the builders dockyard for guarantee maintenance and to install further equipment. The VDS was fitted and also the AGOUTI system to reduce radiated noise. Reducing the acoustic wave of the ship silencing this propagation with an air bubble curtain in the water. (Similar to US PRAIRIE system). Ship not decommissioned but 'Out of Routine' with reduced crew.
On 29 April Commander Crince le Roy assumed command. From 12 - 14 June the 'Dutch Leander' visited Portland meeting old friends and walking the Weymouth Esplanade. From June 23 - 1 July a visit to the Haakonsvern Naval Base Established in 1962 near Bergen, Norway.
28 October: *Van Speijk* departed Den Helder bound for Toulon. With assistance of A.S.M (Anti Sous Marine) staff of the 'Marine Nationale' An evaluation program was promulgated to be carried out in French Exercise Areas, sound range Le Bruse, a fruitful month. Returning to Den Helder on 12 December 1968.

1969
In February/March the Submarine *Zeehond* allocated Sparring Partner for *Van Speijk*. On 28 January *Van Speijk* departed to carry out ASW evaluations. Visited Cadiz and arrived in Gibraltar for a 10-day 'Plan Onderhoud' a rigid new developed maintenance system…

Zeehond also arrived at 'the Rock' and in company with *Van Speijk* visited Tangier, Casablanca, Las Palmas, Malaga., arrival Den Helder 29 March. From 5-16 May ASW evaluations and a visit to Stavanger.

During ASW evaluations
The most sophisticated ASW unit of the RNIN had modern 3-cylinder subs *Potvis*, *Zeehond* and *Tonijn* as adversaries.
30 May: *Van Speijk* in the TG ('Smaldeel 5') as a unit of 'Jagers Divisie 3' (DESDIV 3)

2 June: Londonderry - 15 June COMDESDIV 3 embarked and 19-23 June - Helsingborg; 5 July Summer Leave.

On 25 Aug. *Van Speijk* enjoyed a sailing Family Day in company with *Zeeland* (D 809). From 1-11 Sept: Exercise Knockabout in the Atlantic, *Van Speijk* 'shadowed' a Soviet Kresta class missile cruiser. In Portsmouth 12-15 Sept. and from 17-23 Sept. The NATO Exercise 'Peace Keeper' was carried out, followed 24/25 Sept. by

Multiple RAS with FGS Frankenland (A 1439). To port of the German oiler a Hamburg-class destroyer. (NIMH)

FLAGS AND PENNANTS

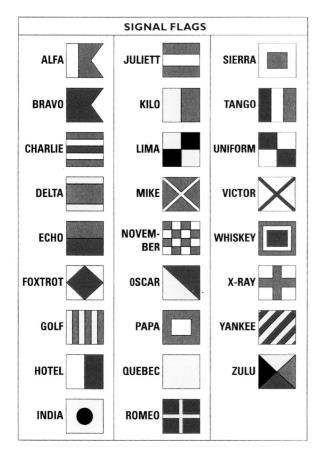

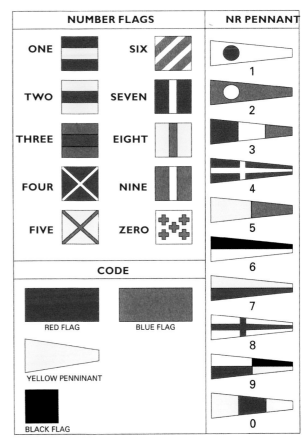

Frigate *Van Speijk*: international callsign PAVA = Allocated by ITU (International Telecommunication Union)

- International callsign also referred to as Signal Letters.

- Visual callsign: F 802 = Promulgated by NATO (ACP 118- ACP 129 NATO sup 1)*

- Encrypted callsign: On tactical voice circuits to disguise identity of transmitting stations (units) Letter/cipher combinations at random e.g.: L6K – 5OB – X0M valid 24 hours.

Note: When radio silence imposed, manoeuvring could be executed by "Flashing Light"
- Range 4/5 nm depending weather.
- Signalling Lantern, diameter 10"(1500 Watt)

Example in a high speed surface action group (SAG)- (Formation Oscar) fast manoeuvring by single letter flashing light signals. (no electromagnetic emissions....).

*Left: Division pennant indicating COM(NL)DESDIV * (Capt. D) embarked..*
"Commandant Jagersdivisie 3"…
The appellation CJAGDIV 3 (COMDESDIV 3) was introduced in the RNLN since 1963.

Hoisting visual callsign the ciphers are displayed by the (international) pennants.

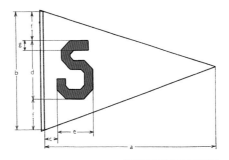

---

* ACP = Allied Communication Publication.

* **Note:** Since 1974 JAGDIV was defined as FREGRON

the PXD (Post Exercise Discussion) at Plymouth. 26 Sept. arrival Den Helder.

28 January: Departure for EXPO 1970 at Osaka. At 1030 'Divisions' for *Van Speijk* and *Van Galen*, the crews mustered on jetty 22/23. The CINCRNLN* would address the officers and ratings and deliver a farewell speech. After 'At Ease' was ordered the

admiral beckoned and invited the men to come closer. A bit hesitating both crews shuffled forward, surrounded him and listened to his informal speech.
At 1200 Family, friends and well wishers were requested to leave the ship.
At 1215 'fore and aft' parties were ordered to their rope handling stations,
At 1230 ships were casted off fore and aft and set course to the English Channel.

WESTWARDS!

The first evening at sea a small ceremony. The CO opened the new PO's mess. Cheers! Encountering some fog, cold and stormy seas, the weather improved when approaching the Azores, arrival Ponta Delgada on 2 February. At home Members of Parliament asked anxious questions about the visit to Curaçao, Indonesia and South Africa.

---

*   Vice Admiral J.B.J.M. Maas (1914-1972) was a protagonist of less rigid communications between superiors and subordinates.

Informal speech before departure. (Coll. H. Visser)

FREGATTEN OP WERELDREIS

Right:
Arriving in Willemstad.
(Coll. H. Visser)

Also about operations of two new A/S frigates outside NATO areas.
Deputy Minister of Defence (Navy), 'Iron Adrian' van Es, dismissed all this noise. The day after passing the Tropic of Cancer dress code: tropical! En route to Curaçao a fine celebration of Carnival; to name Prince Emile I supported by the bands 'Helicats' and 'Hollandse Bieten'. Entering Anna Bay while passing Fort Amsterdam the national anthem was played.

Only 8 months ago (30 May) serious riots in downtown Punda (Willemstad). Police was assisted by marines from Aruba and Curaçao and within 48 hours 300 marines were airlifted from Holland.
From 11 to 18 February the ships were moored at the Brion Quay for a friendly visit to Curaçao. On 18 Feb., there was half a day delay by fuelling in Caracas Bay. To maintain Sailing Scheme with 21 knots to Colon where the ships anchored early

in the morning off Christobal. It was 20 Feb. when locked through the great Gatun Locks to a 90 feet higher water level. In the afternoon the pilot gave wide berth to an oncoming vessel and *Van Speijk* touched lightly the rocky bottom of the Canal causing damage to starboard screw. A quick reaction from the San Diego Marine Construction Co. resulted in a couple of days in drydock in San Diego so the "World Cruise" could be continued.

Passing Panama Canal.
The length of the Panama Canal from shoreline to shoreline is about 40 miles (65 km) and from deep water in the Atlantic to deep water in the Pacific about 50 miles (82 km). The canal, which was completed in August 1914, is one of the two most strategic artificial waterways in the world.

Three days after *Van Galen* departed *Van Speijk,* repairs completed, left San Diego on Sunday 8 March and ordered revolutions for 24 knots (course 252) to make up for time lost and to enter Honolulu in company with *Van Galen.* Successful, arrival at Honolulu Roads 12 March at noon.

The next target was Midway with a large population of Albatross birds (Gooney Birds). Tourists not welcome, only military personnel allowed. Delicate manoeuvring by the pilot negotiating the reefs of the atoll when entering the lagoon.

To reach the important next destination the ships had to sail for one week in the Pacific with wind and sea increasing.

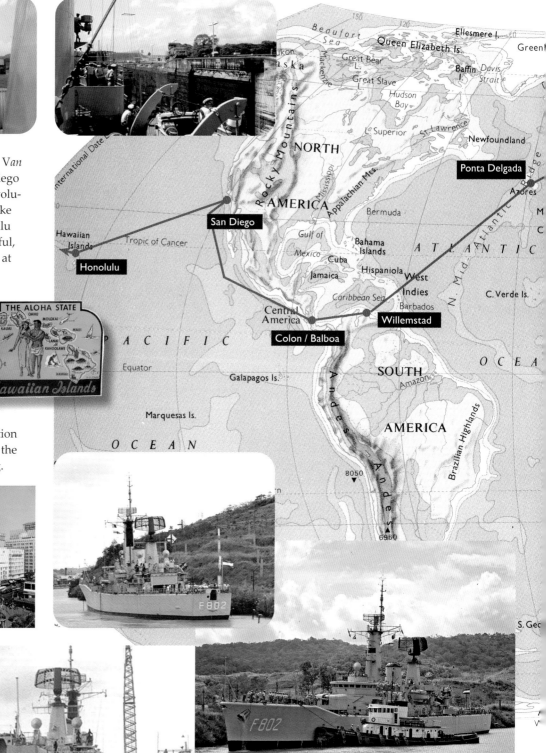

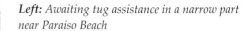

Left: *Awaiting tug assistance in a narrow part near Paraiso Beach*

Right: *Tugboat John F. Stevens on the job.*

Bottom: *Repairs in San Diego*

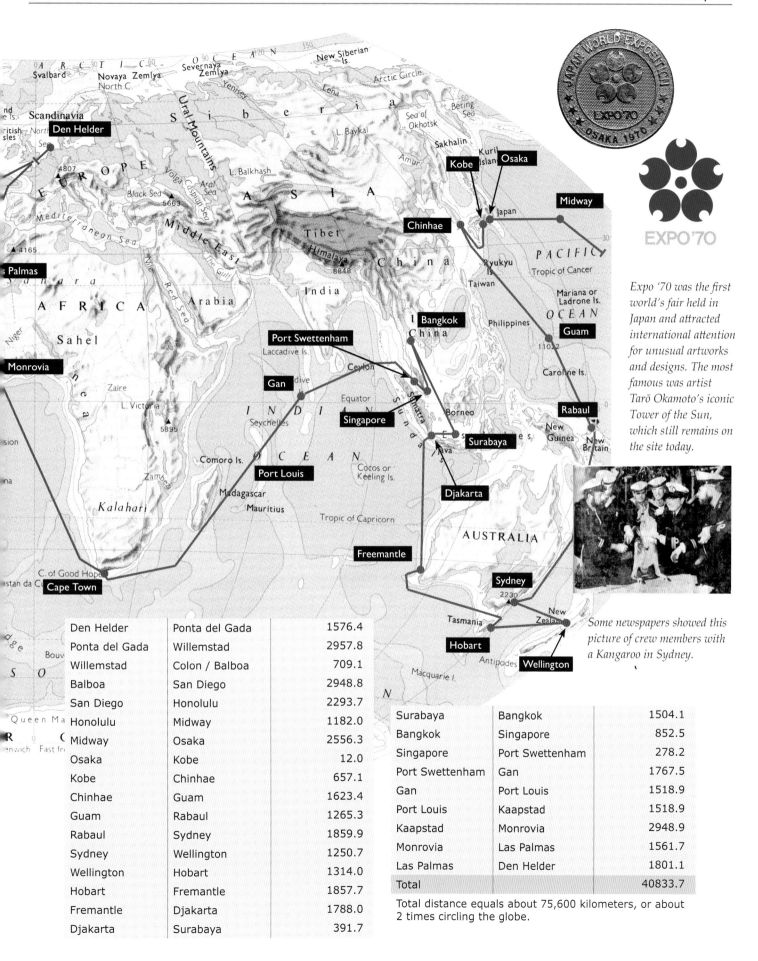

Expo '70 was the first world's fair held in Japan and attracted international attention for unusual artworks and designs. The most famous was artist Tarō Okamoto's iconic Tower of the Sun, which still remains on the site today.

Some newspapers showed this picture of crew members with a Kangaroo in Sydney.

Den Helder	Ponta del Gada	1576.4
Ponta del Gada	Willemstad	2957.8
Willemstad	Colon / Balboa	709.1
Balboa	San Diego	2948.8
San Diego	Honolulu	2293.7
Honolulu	Midway	1182.0
Midway	Osaka	2556.3
Osaka	Kobe	12.0
Kobe	Chinhae	657.1
Chinhae	Guam	1623.4
Guam	Rabaul	1265.3
Rabaul	Sydney	1859.9
Sydney	Wellington	1250.7
Wellington	Hobart	1314.0
Hobart	Fremantle	1857.7
Fremantle	Djakarta	1788.0
Djakarta	Surabaya	391.7

Surabaya	Bangkok	1504.1
Bangkok	Singapore	852.5
Singapore	Port Swettenham	278.2
Port Swettenham	Gan	1767.5
Gan	Port Louis	1518.9
Port Louis	Kaapstad	1518.9
Kaapstad	Monrovia	2948.9
Monrovia	Las Palmas	1561.7
Las Palmas	Den Helder	1801.1
Total		40833.7

Total distance equals about 75,600 kilometers, or about 2 times circling the globe.

Left:
19 March 1970,
refuelling at Midway.
Van Speijk *and*
Van Galen *in the*
main atoll. ESM:
Both ships carried
UA-8 and UA-9 but
Van Speijk *had a*
FH-5 ("bird cage")
antenna in top so
the two masthead ob-
struction lights were
in a lower position.
(Coll. H. Visser)

THE MAMMOTH **NEW OSAKA**

27 March 1970

Sometime after landfall the Expo 70 visitors sighted smart looking and perfect manoeuvring Japanese destroyers *Yudachi* (Thunder shower in summer evening) and *Harusame* (Spring Rain). They escorted the frigates to Osaka Harbour where two tugs spouted streams of coloured water from their fire monitors. A very hearty welcome indeed!

Many crews members visited the Exposition dressed in uniform and sometimes classes of school childeren lined up to ask for a signature… The average age of the crew was about 25 years. The 'Open House' attracted 10,000 visitors and stimulated many friendships.

When leaving on 6 April the quay was crowded with farewell wishers. Some friends were very worried to hear our next port of call would be Chinhae in Korea!

On 6 April the ships departed the naval base of the South Korean Navy. Off Chinhae 8 April, by a ROK patrol vessel the liaison officer and some officials boarded to hand over the national flag of Korea. To be hoisted in a superior position when the salute of 21 rounds was fired. Berthed alongside *Chung Nam* (73) their ship's band played very good Dixieland music!

KOBE PORT SINKO PIER

神戸新港第5突堤S岸壁

TEL: 32-9503

Issued to all liberty
men: "to prevent
Stray Dogs"…
(Coll. H. Visser)

Netherlands Day:
both crews mustered
at the Festival Plaza.
(Coll. H. Visser)

WELCOME

ON BOARD

HNLM FRIGATES

VAN SPEIJK

and VAN GALEN

While underway the ships performed many exercises. Here the helicopter of Van Galen is airborne while Van Speijk is operating her VDS/

Wasp approaching for landing. (NIMH)

The next day a large deputation of both ships went to the UN Memorial Cemetery Tanggok near Pusan (Busan) to pay respect and place a wreath to commemorate the 117 Dutch military fallen in the Korean War. 13 April refuelled at the USN base Guam.

16 April 1970 1132 lt.
Van Speijk Crossed the Line 3 minutes after *Van Galen.* At 1400 hours Father Neptune,

King of the Ocean, and his spouse Salacia (sometimes called 'Neptunia') boarded followed by his suite of 'bears' and 'barbers'… After a cordial receipt by the CO the bears started to round up the 'pollywogs' and drove them to a large canvas bath tub erected on the heli deck. Many 'pollywogs' were baptized and entitled to call themselves 'Shellbacks'. The next refuel was in Rabaul, New Britain. A message received to monitor some frequencies and to maintain silence on some others. The reason for this was not to interfere with the oncoming landing of the lunar module Apollo XIII which had suffered damage. They landed later in the South Pacific 18 Aug 0400 lt, some 1800 nm from our position. About 2000 nm

to Sydney for the 1770-1970 Bicentenary of Captain Cook! Also the 2000th deck landing by the helicopter of *Van Speijk* had been executed. On 20 April the engine of the Wasp was rejected and had to be replaced by the second reserve engine.

23 April
After passing the Middlehead Rock a salute of 21 rounds was presented before entering Port Jackson and turning to take a berth at Garden Island, the naval base. The ships would take part in the celebrations in memory of the landing of Captain James Cook in Botany Bay in 1770, Also 45 other ships would visit the bicentenary celebrations. In and around Sydney lived about 50,000 'Dutchies'.

As the first real 'shipboard helicopter' the Wasp proved to be sufficient for its task, although the performance in warm weather did not last. In those conditions they often flew without doors and back seats to make the helicopter as light as possible.

Bottom: *Since the hangar had limited space the Wasp, here on* Van Galen, *had to be folded and turned around* (NIMH)

The visit included a maintenance period and would last until 11 May. The crews received an overwhelming welcome, e.g free tickets for the musical Hair, free ride in buses, 'dial a sailor'-telephone line by Wrens for invitations.

> ## Overloaded program
> Example:
> 28 April Parade in the city
> 29 April Reanactment Botany Bay in attendance Queen Elisabeth II and Prince Philip, Duke of Edinburgh
> 30 April Birthday Queen Juliana of the Netherlands.

11 May
Great visit to Australia ended. In the forenoon watch the ships proceeded with HMAS *Perth* (DDG 38) leading to an exercise area and to execute a CASEX with the submarine HMAS *Ovens* (70). *Perth* launched two Ikara ASW missiles. Another country loved by the Dutch is New Zealand. On 14 May the beautiful bay of Wellington was entered with 22 kts. The ships were allowed to berth alongside the Overseas Passenger Terminal on the Clyde Quay Wharf. The hospitality of the 'Kiwi's' matched that of the 'Aussies' culminating in a smashing dance with the two ships' bands presented by the Dutch community in the Esso Motor Hotel! Well done!

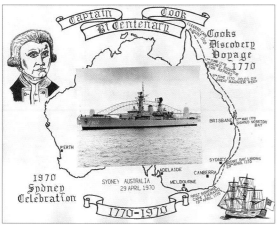

Memorial service in Java Sea.

There was an old Sverdlov cruiser named *Irian* and on frigates the crew stood at attention a smart welcome ceremony.

Commemoration Battle of the Java Sea (27 Feb. 1942).
After departure on the afternoon of 9 June the ships formed line abreast, distance 90 ft. The broadcasts were connected, crews mustered in ceremonial white kit (dress 12). Position 06-00 S 112-05 E of the light cruiser *De Ruyter*. The RADM delivered a wreath and the 'Last Post' sounded by the hornblower. Cleaning ship and take a break the time to reach Bangkok, distance 1500 nm, was 4 days.

Maybe 1350 nm to the South West this happening was intercepted by the Dutch cummunity in Hobart, Tasmania. Many of these Dutch were apple growers and they organized a terrific dance party during the evening the ships were in port.
The last harbour to visit 'down yonder' was Fremantle near Perth and the 500 Dutchies could finally confirm their favourable opinion about the 'Southern Continent'!
In Fremantle RADM J.C.H. van den Bergh embarked, his command flag was hoisted on 30 May (until 12 June). His rank emphasized the importance of the planned visit to Indonesia.

The 3rd June, after a salute to the Red and White flag of the Republic and one for The Naval Commander Jakarta the ships berthed in Tanjung Priok the harbour of Jakarta. (Temp. 36°) Later ceremonies were performed at Kalibata and Menteng Pulo. The people in the streets were remarkable friendly, many wanted to exercise their knowledge of the Dutch language with the "Orang Belanda" (Dutch).

The visit to Indonesia was a successful "Holland Promotion" operation. Many of the ship's complements had Indonesian roots and would meet their family. Communication with the inhabitants, even in a big city like Jakarta (4 M.!), was cordial without a single incident, bearing in mind it was only eight years after the troubles in New Guinea.
The 500 'ambassadors' paved the way for the Royal Visit in 1971. Leaving Jakarta the 5th; after one day sailing the ships were surprised in the morning of 6th June by three Soviet built Komar missile boats; an

escort to Surabaya Madura Quay (former 'Kruiser Kade').
A salute of 13 rounds for the local naval commander. Deputations of both ships partook in ceremonies at 'Fields of Honour' (Military Cemeteries) Kusuma Bangsa and Kembang Kuning.

Indonesia–Netherlands relations
Indonesia and the Netherlands established diplomatic relations after the independance in 1949. (Indonesia was the largest former Netherlands colony.) In 1956, the government of Indonesia, led by Sukarno, cut off all diplomatic ties with the Netherlands, ties that were restored only in 1968 by the New Order government. In May 1970 the two frigates visited Indonesia. In return Indonesian President Suharto in August 1970 paid an official visit to the Netherlands, which was reciprocated by Dutch Queen Juliana and Prince Bernhard royal visit to Indonesia in 1971.

Indonesia
- Largest Netherlands Colony for ±300 years.
- From 1945 "War of Independence" President Sukarno.
- 1947-1948 = 'Politionele Acties' = 'Law and Order Enforcements'.
- 6600 men NL Forces rest in various Fields of Honour, Cemeteries; kept very well. The Republic was recognised in 1949.

13 June. Arrival at Bangkok in a monsterous heat (40° C), this restricted the number of libertymen. In the morning of 19 June the ships berthed at Singapore. The next day they anchored in the Men of War anchorage, later shifting berth to the dockyard at the naval base HMS *Terror* to carry out regular maintenance.

Departure 26 June to Port Swettenham, the harbour of Kuala Lumpur. A refuelling point in the Indian Ocean was the island of Gan (Addu Atol); residence tanker RFA *Wave Victor* (A 220). A visit to the island Mauritius followed, independent since two years. Underway to Cape Town *Van Speijk* steamed with economical speed, there was some ocean swell but the focsle kept dry so two quartermasters (PO's) were busy to check anchor gear. Suddenly, pitching in a head sea the ship was hit by a slammer

sweeping the PO's off their feet. One suffered injuries (broken bones), the other only a light concussion.

17 July: 'Land Ho!' Table Mountain dominated the panorama of Cape Town. Libertymen often were addressed in 'Afrikaans', a language with deep Dutch roots. This country was bilingual! A fine four-day visit! much of the programme organised by the Dutch Ambassador (Baron Lewe van Aduard). This visit was watched by a critical press. Only some minor incidents regarding refusing admittance.

Unrestricted admittance for the crews if dressed in uniform had been assured but some owners of restaurants (or visitors) did not comply.

SEAMEX 4 (L) = Light Jackstay Transfer.

21 July. Departure from beautiful Cape Town.

26 July. En route to Monrovia R/V with RFA *Wave Chief* (A 265) to execute a RAS.

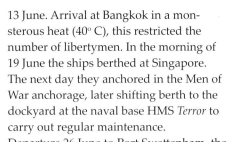

STANAVFORLANT, Van Speijk *sequence nr 2 in column (NIMH)*

On 29 July the ships performed a ceremonial entry and fired a salute of 21 rounds to the flag of Liberia. Berthed on the L.M.C. Quay Monrovia. The next day President W.V.S. Tubman, head of state since 1944, would like to pay a visit to H.M. ships. He was a good friend of Queen Juliana! About noon at least 12 motorbikes dashed to the quay followed by a column of caddish cars packed with security guards some posted on the footboards of the cars. The showy happening stopped in front of the gangway.

3 August. The ships were at Las Palmas berthed alongside the long mole to refuel for the last leg home and proceeded with a rather high speed of advance (21 kts). Saturday 8 August in the early hours of the Middle Watch anchored in 'Marsdiep' (Den Helder Roads). After embarking the CINC and customs entering the homeport in the rain to berth at jetty 19 to be welcomed by 1000+ relatives who despite the rain broke through the security cordon. After 192 days and 40,833.7 nm: mission completed.

NL Task Group 1971
Starting 1971, *Van Speijk* was assigned to the NLTG but still in Den Helder for maintenance. On 8 January she sailed for individual exercises in the North Sea, English Channel and Brest areas. Visiting Brest (15 - 18 Jan), before returning 22 January. By 5 February, the NLTG was divided in Task Units, a new chain of command following changed NATO guidelines. Being assigned to Task Unit 429.5.1 (*Van Speijk* and ASW destroyers *Rotterdam*, *Amsterdam* and *Gelderland*) perform exercise Creaky Nut in the North Sea. On 27 February departing for exercises near

Land's End and a visit to Portsmouth. Followed by exercises in the English Channel and Bay of Biscay. Returning 12 March for some maintenance.
While the other ships of the TU visited Amsterdam, *Van Speijk* paid an informal visit to Dublin (16 - 19 April) after which the TU was reassembled in the North Sea. On 24 May she was released from the Task Unit for scheduled maintenance.

1972
In summer the ship was reactivated and made some short journeys for testing

and evaluating. To calibrate the Limbo mortar, she visited Portland (5 - 7 Sep) and sailed to the Irish Sea, north of Scotland to Norway conducting speed trials[**] in the Björnefjord followed by VDS trials in the Hardangerfjord. The weekend was spent in Trondheim (15 - 18 Sep). Returned to Den Helder (21 Sep) for some minor repairs. On 27 October assigned to the NL Task Group and joined exercises (30 Oct - 10 Nov), visited Lisbon (4 - 5 Nov). There was hardly enough space to berth the ships. Anchoring in front of the Doca de Marinha, became a little adventure when strong current in the Tagus dragged the anchor. On 10 November *Van Speijk* left the TG and headed for Den Helder for improved adjustment of automatic boiler control and prepare for FOST (Flag Officer Sea Training) in Portland (24 Jan - 16 Mar 1973).

1973
Van Speijk was assigned to the NLTG (2 Apr - 29 Jun) and present at the national combined exercise Rocking Nut (14 - 25 May) in the North Sea and Norwegian Sea. Visiting Trondheim (18 - 19 May).
From 18 July to 19 September in Standing Naval Force Atlantic.

** Measured mile

New York Area
1609 – early settlers. Colonists in 1624. Provincie Nieuw Nederland and citadel Fort Amsterdam. Nieuw Amsterdam founded in 1626; City rights in 1653. Named New York in 1664, confirmed in 1674 (Westminster Treaty).

Towing practice with ASW destroyer Friesland.
Seamex 2 (FOST Guide) In good weather: 6 knots made good is realistic. (NIMH)

1974
On 7 January V*an Speijk* joined NLTG and with *Holland* crossed the Atlantic to perform escort duties, screening the main body during NATO exercise Safe Pass (5 - 14 Mar). Homeward bound the ships exercised with STANAVFORLANT (18 - 25 Mar) and FOF1 (flag officer first flotilla). After exercises the two Dutch units sailed independently. Visiting Ponta Delgada, Bermuda (Ireland Island) and Norfolk. Dismissed on 13 May for work-up at FOST (20 May - 11 Jul). On 30 August she was once more assigned to the NLTG. The Task Group visited Helsinki (2 - 7 Sep) for showing the flag during the visit of the Queen and spouse to Finland. Visiting

Glasgow (16 - 27 Sep) before joining major NATO exercise Northern Merger (16 - 27 Sep). Returning to Den Helder on 29 November.

1975
Still with the NLTG. On 27 January the group sailed to the Gibraltar area. En route the NLTG performed exercises with British- and French aircraft in Dover Straits. Also spotted a submerged Soviet Juliett Class submarine accompanied by a merchant. Visiting Gibraltar (7 Feb) before commencing exercises with British units (10 Feb) and submarine *Tonijn* (11 - 13 Feb). A visit to Casablanca (14 - 19 Feb) was used for some minor maintenance.

Also, Brest was visited (2 Mar) where the Minister of Defense embarked *Van Speijk* to witness the fleet activities while doing exercise Hot Dog (3 - 4 Mar) with French and Dutch aircraft. On 5 March the ship's helicopter ferried the VIP to Vlissingen.

On 15 May *Van Speijk* and *Van Galen* departed for an 8 week journey to the United States and Canada. While proceeding to Norfolk via Ponta Delgada (20 May) and Bermuda (26 - 28 May) performing several exercises. On 21 - 22 May joined by Orion MPA's of the US Navy's Naval Air Facility Lajes (NAF Lajes) (Azores). Visiting Norfolk (30 May - 2 Jun and 6 - 18 Jun) and conducting trials with US units. Both frigates went to New York (19 - 24 Jun) in the context of *'A salute of the Netherlands to New York'* to commemorate the founding of New Amsterdam about

350 years ago. Among the festivities was a gathering at Battery Park where 40 of the crew re-enacted the landing of the first settlers.

After departure (24 June) the frigates had a rendez-vous with large missile destroyer USS *Macdonough* (DLG-8/DDG-39) for exercises during the day. Now heading for Halifax (24 - 30 Jun) and St. John (2 Jul) for replenishing. The frigates departed some hours later. On 5 June, while proceeding exercises with Orion MPA's of the US Navy. Arriving in the English Channel on 8 June to act as sparring partner for units conducting FOST while at the same time an air defence exercise was in progress. The ships returned to Den Helder on 10 July.
On 20 October, *Van Speijk* and ASW destroyers *Overijssel* and *Limburg* sailed for the Mediterranean to participate in NATO exercise Iles d'Or (3-18 Nov) involved *Van Speijk*. While underway GUNEX (gunnery exercises) were conducted and some with French units.
On 24 October *Van Speijk* assisted a burning Spanish fishing vessel and rescued 14 of the crew. Near St. Croix the men were transferred to a tugboat. Continuing the journey to the point of rendez-vous with County-class destroyer HMS *Hampshire* (D-06). Followed by exercises with aircraft of USS *Independence* (CVL-62) (27 Oct), a short visit to Gibraltar (28 Oct) and exercises with US MPA's. On 31 October the ships arrived at Toulon to prepare for the NATO exercise involving 30 ships from 6 countries. Returning on 26 November.

1976
For the winter cruise the TG departed 26 January. Units: *Evertsen, Van Speijk, Zuiderkruis, Limburg* and *Friesland.* Exercises area west of Gibraltar. *Tromp* joined 9 February. Exercise Spring Train. *Van Speijk* (CFREGRON embarked) with *Evertsen* and *Zuiderkruis* departed Gibraltar (18 Feb) in convoy with RN units (HMS *Ark Royal* (R-09), RFA *Lyness* (A-339) and RFA *Olmeda* (A-124), proceeding to Roosevelt Roads to prepare for NATO exercise Safe Pass (8 – 19 Mar). Although the exercise was in bad weather conditions near the US east coast, it ended with the PXD (Post Exercise Discussion) at Halifax (9 – 22

Mar). Departing homeward on 23 March in company with *Evertsen.* The icy conditions forced the ships to sail a more southern course when they ran into a dense fog. Needless to say, the crew did not see any icebergs. Arriving 2 April.

After some smaller exercises *Van Speijk* and *Limburg* paid a visit to Rosyth (14 - 17 May) and participated in JMC762 (Joint Maritime Course) until 28 May. *Van Speijk* returned to Den Helder while *Limburg* paid another visit to Rosyth.
From 9 to 20 August *Van Speijk* executed trials in the North Sea and visited Swansea (13 - 16 Aug). By the end of the month, again assigned to the NLTG, now with *Tromp* (flag), *De Ruyter, Poolster, Van Nes* and *Isaac Sweers.* On 31 August the ships hosted politicians who reviewed navy life

for a day. The guests were divided over the units. For the occasion frigate *Van Galen* was added to the TG. In the following days the TG performed exercises in the North Sea. On 15 October *Van Speijk* was involved in the search for survivors after a distress call of the German coaster *Antje Oltmann.* On her way from Rotterdam to Cork she ran into trouble when the cargo shifted. The search party rescued 3 and recovered 5 of the crew of 9.

On 24 December *Van Speijk* was decommissioned and relocated to the Naval Shipyard for her Mid Life Modernisation (MLM). The ship underwent an array of modifications. Plagued by time-consuming start-up problems it would take about two years. Price tag; about 60 million Dutch guilders each ship.

Post MLM

On 3 January 1979 *Van Speijk* was commissioned and started trials. In the first months a new list of modifications emerged. Some of these were carried out later in the year while in service. While the other ships received MLM they were converted following the new data. *Van Speijk* was now sailing with reduced crew (70 less) so the ship's management had to be adjusted. On 22 January *Van Speijk* departed for her first operational journey in the North Sea. Visiting Plymouth (23 - 26 Jan), Lisbon (2 Mar) and Gibraltar (5 – 7 Mar). On return the ship saved 4 crew members of the German coaster *Jasmin* (11 Mar) in the Bay of Biscay. Arriving in Den Helder 14 March.

On 22 October she sailed to establishing wind speed limits for operating the SH-14B Lynx ships helicopter. A team of specialists of the National Air and Space laboratory were embarked for test and evaluation. Visited Hamburg (2 – 5 Nov). On 11 November she made a rendezvous with hydrographical research vessel *Tydeman* (A 906) which suffered engine failure. After establishing a line, *Van Speijk* tugged *Tydeman* to Den Helder.

On 12 November *Van Speijk* received regular 8-week maintenance.

Reconstruction of Van Speijk class, here Van Galen *at the Royal Shipyard in Den Helder. The work is by far not completed, note the man in the topmast working at the smaller antennas.*

STANAVFORLANT

In de first 6 months of 1980 *Van Speijk* joined Standing Naval Force Atlantic. Entering Portsmouth on 9 January where the squadron (HMS *Ardent* (F-184), BNS *Westdiep* (F-911) and FGS *Schleswig-Holstein* (D-182)) was assembled. Two days later the ships left port heading for Ponta Delgada. A short stop was made for replenishment at the Azores (15 Jan) before crossing the Atlantic. The splendid weather conditions offered good opportunities for exercising procedures and streaming the VDS (Variable Depth Sonar). Arriving at Bermuda (21 Jan) for a 2-day visit. The next day bad weather forecast urged the ships

Approaching port of Den Helder.

to leave early in the afternoon. In Mayport HMCS *Skeena* (DDH-207) and USS *Luce* (DDG-38) completed the squadron. The Squadron commander was embarked in the American ship. On 29 January the Squadron departed for work up and headed for Savannah leaving the warm Gulf Stream and meeting less comfortable weather. It was hurricane season off the Florida coast. The ships received a warm welcome in the port of Savannah. The highlight for *Van Speijk* was winning the 'Culinary Exhibition' in the Hilton.

From 18 to 22 February the squadron concentrated on ASW capabilities by executing a series exercises before entering Mayport to prepare for the upcoming NATO exercise Safe Pass 80. About 40 ships contributed and it was being conducted in the Bahamas at the Andros Missile Firing range*, en-route to Nova Scotian waters. Unfortunately, running into rough weather. Once the exercise commenced (25 Feb) the weather deterio-

Right:
STANAVFORLANT March 1980: top to bottom, USS Sellers (DDG11), Schleswig-Holstein (D182), HMS Juno (F52), HMCS Annapolis (265), Hr.Ms. Van Speijk (F802) and BNS Westdiep F911

---

*   AUTEC = Atlantic Undersea Test & Evaluation Center

rated quickly and for two days and nights, the ships were pounded and lashed by 80 Mph gale and broken water. The ships were buffeted by green seas. Some washed completely over the bridge. ENDEX before entering Halifax (7 Mar) for debriefing and thereafter to Charleston for a scheduled 3-week maintenance.

Charleston, March 1980. Command change of SNFL: From Capt. G.M. Carter USN to Capt. D.G. D'Armytage RN.

Many relatives took the opportunity to spend holidays with their beloved. Also, three new NATO ships were welcomed: USS *Sellers* (DDG-11), HMS *Juno* (F-52) and HMCS *Annapolis* (DDH-265).

On Easter Monday (7 Apr) the squadron departed. While working up, 4 Sea Cat missiles were launched, and the Oto Melara gun was intensively used. Reaching HMS *Malabar* at Bermuda on 14 April for a 2-day visit before returning to European waters. A rendez-vous with the new fast combat support ship *Zuiderkruis* was made on 21 April.

On 24 April exercise Open Gate commenced in Lisbon areas. Shortly after BNS *Westdiep* was dismissed while FGS *Schleswig-Holstein* was released by FGS *Emden* (F-221). Departing Lisbon on 6 May. Passing Strait of Gibraltar into Mediterranean for exercise Dawn Patrol. While the exercise was in progress the squadron anchored in Augusta Bay for a briefing. When preparing to leave, USS *Sellers* grounded. An investigation was carried by divers of *Van Speijk*.

On 15 May one of the crew fell overboard. Within 7.5 minutes the unfortunate sailor was back on the deck! After ending the exercise (17 May) the ships went to Naples. On 26 May the squadron departed for a last mutual exercise. On 3 June the squadron was disbanded and a steampast carried out. Arriving Den Helder 4 June.

On 5 and 8 June *Van Speijk* went for some hours in the North Sea hosting some NATO ambassadors.

On 18 August departing in company with *Zuiderkruis* and *Zwaardvis* for exercises. Visiting Göteborg (22 - 25 Aug) and taking up duties as ready duty ship in the North Sea. After release replenishing at Haaksonvern (1 Sep). Departing the following day for torpedo trials in the Björnefjord with *Zwaardvis*. Visiting Bergen (4 - 7 Sep) before heading to the Rona Range in British waters for VDS trials. Replenishing at Rosyth (12 - 15 Sep) and taking up guard ship duties for a week. On 22 September *Van Speijk* returned to Den Helder. A major overhaul (two-yearly) followed.

1981

In spring 1981 *Van Speijk* joined Task Group 429.5 with: *Tromp* (flag), *Zuiderkruis*, *Van Galen* and *Banckert*. Departing 21 April for work up, followed (25 Apr) by a series of ASW exercises in territorial waters. It was within range of the air force which contributed by several simulated attacks (JANEX) and deployment of target drones. The ships anchored in the weekend off the Danish coast before heading to the Skagerrak exercise areas. Again ASW exercises with a 'hostile' Danish submarine; assistance of British Nimrod MPA's. Some variety by a GUNEX. On 30 April the TG 429.5 visited Copenhagen. On 4 May the ships sailed to the south coast of Norway. Off Bergen a series of exercises were held with *Dolfijn*, British Nimrod (from Kinloss) and US Orion's

RAS: Zuiderkruis *transferring stores and liquids to* Van Speijk *(NIMH)*

Enlarged living spaces after MLM (NIMH)

Helm relocated to bridge. (NIMH)

(from Keflavik) MPA's. Visiting Bergen (8 to 11 May) where emerging rudder problems turned out to be more serious than expected. *Van Speijk* was ordered to Den Helder for repairs.

10 – 21 August busy with exercises and wandering around

Seamanship practice training for a class of OPS ROOM officers. Commenced cheerfully on Monday with the standard mist navigation track through the narrows of Schulpengat near Den Helder. The training areas were the North Sea, Thames estuary, English Channel, to the Scilly Isles and exercising near/in the Solent, Wight, Portland, Brighton, Swanage and Poole Bay; a visit to St. Malo and training off Vlissingen, Scheveningen and Rotterdam! Sparring partners: *Jaguar*, later *Banckert*. Ship handling:

- pick up DAN-buoy/swimmer
- lower boat / launch dinghy
- man overboard and pick up by Williamson turn or full astern?
- anchoring manoeuvres (Brighton)
- mooring (Portland)
- blind pilotage in shallow water/swept channel
- towing
- RAS approaches
- emergency breakaway procedures
- emergency steering (by hand)
- total steam failure

"to mention some lessons to learn"…

Off the south coast of Britain remarkable floating fields of orange-brown plankton were observed.

After summer leave *Van Speijk* returned to the TG 429.5, now with *Tromp* (flag), *Zuiderkruis*, *Van Nes* and *Callenburgh* leaving on 26 August to Portsmouth and Plymouth to attend the British Navy Days. Once in British waters the task group split up. *Tromp*, *Zuiderkruis* and *Van Speijk* sailed to Portsmouth to act as host for visitors. On 1 September the TG was reassembled in the Plymouth Sound and exercise areas.

Centre, left and right:
Action Stations: Bridge of Isaac Sweers 1984
- *Helmet*
- *Anti Flash Mask*
- *Life Jacket*
- *Anti Flash Gloves*
- *Gas Mask*
- *Binocular (Chief Yeoman)*
- *Trouser legs tucked in socks*

*RAS operations
While* Van Speijk *is receiving loads in alongside replenishment, a GW frigate is being refuelled astern by Zuiderkruis.*

*The NATO abbreviation MISCEX means Miscellaneous Exercises (AXP-3 refers) MISCEX 805 followed by letter
(F) =Fuel
(S) = Solids
(L) = Light Jackstay*
(NIMH)

Also HMS *Ambuscade* and *Glamorgan* joined. After the exercise the TG was split up again. Now *Tromp*, *Zuiderkruis* and *Van Speijk* visited Plymouth (4 - 7 Sep), while the others visited Portsmouth.
On 7 September both groups sailed to their starting positions for NATO exercise Ocean Safari 81. This year the NLTG was coded orange, which was a limited role, without any room for initiative. After acting in this role, the NLTG withdrew (14 Sep). On 17 September the ships exercised with Belgian air force and returned to Den Helder.

On 19 October the group departed for the Mediterranean. In the first week the changeable weather with fog and wind force 7 to 8 made the journey far from comfortable. Nevertheless, the exercises continued and the French Super Etendarts

Navy Days Den Helder, 1984. Evertsen (F 815), Van Galen (F 803) and Van Speijk. .
Note the new hood on funnel and the differences in the superstructure at the stern. (Coll. Jt. Mulder)

proved to be serious opponents. Proceeding south the weather cleared and Lisbon (23 - 26 Oct.) was reached in sunny weather.

In the second week submarine *Tijgerhaai* came to the scene and acted as opponent. While British Nimrods and US Orions assisted in the hunt. Over the weekend *Van Speijk* visited El Ferrol. On 2 November the NLTG sailed in dense fog to the Bay of Biscay exercise areas. Now the submarine *Zeehond* was the adversary. After the exercises the ships headed for Plymouth for a weekend visit.

9-10-11 November in Plymouth exercise areas with *Tijgerhaai* and *Zeehond* for CASEX. On 13 November return of the TG.

Winter journey 1982

Ships formed part of the TG: *Tromp* (flag), *Van Speijk, Callenburgh, Piet Heyn, Wandelaar* (BE), *Overijssel* and *Zuiderkruis*. Departing for the United States (8 Feb.) to celebrate the 200 years diplomatic relations between the USA and the Netherlands. While crossing the Atlantic the weather rapidly deteriorated. On 12 February, in the Azores area a distress call was intercepted. Transmitted by the Greek tanker *Victoria*, with a cargo of 22,000 ton molasses on her way from Florida to Liverpool. *Van Speijk* and *Callenburgh* both carrying helicopters were dispatched, despite the distance of 400 nm in poor weather to the rescue. Arriving at dawn to find the still floating aft part of the

merchant pitching to 7 metres (23 ft) and rolling up to 45°. High wind 30 kts - Beaufort 7. Both helicopters rescued a total of 16 crew; 16 others did not survive. Course was set to Ponta Delgada to disembark the survivors. On 21 February *Van Speijk* re-joined the TG.

Several exercises were conducted in areas near Roosevelt Roads where submarine *Zeehond* joined to act as opponent. Afterwards the ships headed for Florida to visit Fort Lauderdale (5 - 8 Mar.) and Norfolk (19 - 29 Mar. and 11 - 12 Apr.).

On 14 April *Zuiderkruis* and *Van Speijk* paid a visit to Philadelphia. The NLTG gathered again on 20 April to proceed to New York to add lustre to the visit of HM Queen Beatrix and Prince Claus to the USA. After a warm welcome in New York Harbor by spouting tugs, the ships moored on 22 April.

Some days later the royal couple visited the Task Group. Leaving on 27 April for exercises with Canadian units and a visit to Halifax (6 - 10 May) before returning to Den Helder.

After summer leave and taking part in exercises Silver Nut (30 Aug. - 3 Sep.) and Northern Wedding (6 - 17 Sep.). Visiting Aalborg (27 - 30 Aug.) and Wilhelmshaven (3 - 6 Sep.).

On 25 October the NLTG departed for a training and exercise journey in the Mediterranean. Notable was that for the first time in history exercises were held with units of the Spanish navy; the frigate *Cataluna* (F-73) and submarine *Tonino* (S 62).

Visiting Lisbon (5 – 8 Nov). *De Ruyter* was directed to Den Helder to make arrangements for the passage to London by Queen Beatrix and Prince Claus for a state visit to the UK. *De Ruyter* and *Van Speijk* were berthed alongside HMS *Belfast* (16 - 22 Nov). In the heart of London, *Van Speijk*, the Dutch *Leander* showing Dutch craftmanship on British design. While the others were anchored off Greenwich. On 23 November the ships returned for celebrating 200 years Naval Base Den Helder.

1983

On 31 January the NLTG (*De Ruyter* (flag), *Zuiderkruis*, *Piet Heyn*, *Van Kinsbergen*, *Bloys van Treslong*, *Van Speijk* and Belgian frigate *Westhinder*) departed for the exercises Ginger Nut (21 Jan - 4 Feb) and Roebuck 83 (4 - 11 Feb). When Roebuck 83 commenced *Van Speijk* was still in Haakonsvern for machinery repairs, joining 8 February. Debriefing was at Rosyth.

On 15 February *Van Speijk* returned to Den Helder and was withdrawn from the TG and made preparations to act as guard ship in the West. From 9-11 July a visit to Kristiansand. On 24 November she relieved *Van Nes* at St. Maarten.

1982, visiting London.
(NIMH)

1984

Visiting Kingston (14 - 16 Apr), Key West (18 - 21 May) and New Orleans (24 - 28 May). After seven months torn was completed, arrival in Den Helder on 25 June. In FREGRON (24 Sep - 20 Oct), with *Banckert*, *Van Speijk* visiting Copenhagen (5 - 10 Oct) and Helsinki (12 - 15 Oct).

1985

Van Speijk received maintenance combined with repairs of the boilers while higher command prepared a possible transfer to Indonesia. On 13 September *Van Speijk* was decommissioned and stricken from the list.

1986

On 10 February, Indonesia and the Netherlands signed an agreement for transfer of two *Van Speijk* class, with an option for two more ships. *Van Speijk* was transferred on 1 November 1986 and renamed KRI *Slamet Riyadi (352)*. The ships were provided with all spare parts but no helicopters. The Royal Neth. Navy was responsible for initial training of the Indonesian crew. Now a unit of *Ahmad Yani* class although many Indonesians kept referring to the ships as *Van Speijk* class. (In 1989 it was announced that Indonesia would also obtain the last two frigates.)

Lowering the ensign for the last time. (NIMH)

By 2002, the ship's Seacat missiles were inoperable, and it was reported that propulsion problems were badly affecting the availability of the ships. *Slamet Riyadi* was the first to be modernised by PT Tesco Indomaritim. It was completed in 2008. The ship's Seacat launchers were replaced by two Simbad twin launchers for Mistral anti-aircraft missiles, and she was re-engined with two 10.9 MW (14,600 shp) Caterpillar 3616 diesel engines. Trading in speed for better acceleration and more economical fuel consumption. As the Indonesian Navy retired Harpoon missile from its stockpiles, *Slamet Riyadi* was rearmed with Chinese C-802 missiles.

Series editor	Contributors
Jantinus Mulder	NIMH
	Dick Vries
Publisher	Michiel Woort
Walburg Pers / Lanasta	
	Graphic design
Authors	Jantinus Mulder
Jantinus Mulder	
Henk Visser (FCCY rtd)	**Corrections**
	Crius Group

First print, January 2023
ISBN 978-94-6456-056-5
e-ISBN 978-94-6456-057-2.
NUR 465

Contact Warship
jantinusmulder@walburgpers.nl

Lanasta

© 2023 Walburg Pers / Lanasta

www.lanasta.com

Slamet Riyadi was decommissioned on 16 August 2019. After decommissioning, the OTO Melara 76 mm gun was landed for naval gunnery training at the range in Paiton, Probolinggo Regency, East Java. In 2020, the ship's hulk was sunk offshore of the Karangasem Regency, Bali as dive attraction.

References

- Bremer, F.O.J. Ir. *Radar Development in the Netherlands.* Thales 2004.
- Chesneau, R. *All the world's fighting ships .* Conway Maritime Press 1980.
- Friedman, N. *The postwar naval revolution.* Conway maritime Press 1986.
- Marriott, L. , *RN Frigates since 1945.* (2nd ed) Ian Allan Ltd. 1990.
- Meyer, C.J., Cdr. OBE RN, *Modern Combat Ships 1: Leander Class.* Ian Allan Ltd. 1984.
- Mulder, Jt., *Warship 02: Frigate HMS Leander.* Lanasta, 2013.
- Nooteboom, S.G., *Deugdelijke Schepen.* Europese Bibliotheek 2001.
- Osborne, R. & Sowdon, D., *Leander class.* World Ship Society 1990.
- Romijn, D.E.D. Ir., *De fregatten van de Van Speijk klasse.* Marineblad Jan. 1968.
- Visser, H., *Foto's & Feiten 1950-1975.* Lanasta 2010.
- Bogers, M.J. KTZT (Confidentieel) Report: *Project MLM Van Speijk* 1975
- Voort, T.J. v.d. LTZ 2OC *Wereldreis 1970* (VBZ Union Magazine, 1970)
- *Jaarboek van de Koninklijke Marine* (various volumes)
- *Alle Hens: Tweede jeugd voor Van Speijk fregatten, January 1979:*
- *Warship World* (various issues)
- Website: *Dutchfleet.nl*
- *Zeewezen April 1965.*